Arching in Piled Embankments Including the Effects of Reinforcement and Subsoil

Yan Zhuang

Arching in Piled Embankments Including the Effects of Reinforcement and Subsoil

 Springer

Yan Zhuang
School of Civil Engineering
Southeast University
Nanjing, Jiangsu, China

ISBN 978-981-96-1197-3   ISBN 978-981-96-1198-0 (eBook)
https://doi.org/10.1007/978-981-96-1198-0

This work was supported by National Science Foundation for Excellent Young Scholars of China (51922029) and National Natural Science Foundation for General Program of China (52178316).

Preface

The stability of an embankment constructed on soft soil is governed mostly by the shearing resistance of the foundation, and the construction of an embankment on soft soil may be limited to a problem of bearing capacity (i.e. serviceability or ultimate limit state conditions of the proposed embankment). Piled embankments provide an economic and effective way to conquer the problem of constructing embankments over soft soils. Yet in reality, this technique has already been used many times over throughout history for building on difficult sites, without necessarily a solid understanding of its mechanism and behavior.

The performance of piled embankments is highly controlled by the ability of the granular embankment material to arch over the 'gaps' between the adjacent pile caps. Geogrid or geotextile reinforcement may be used to enhance the resistance of embankments to avoid failure through excessive deformation or shear in the foundation, although the role played by reinforcement has not been fully understood so far. Design methods for piled embankment are conventionally proposed to estimate the stress (after arching) acting on the underlying soft ground which remains independent on the properties of the soft ground. The stress is then generally used to determine the amount of geogrid or geotextile reinforcement required in the embankment system. However, the estimation of stress after arching may vary considerably under the different design methods.

The predominant objective of this monograph is to examine the performance of the reinforced piled embankments in both plane strain and three-dimensional conditions. The arching mechanism involved in the piled embankment has been investigated by conducting a series of finite element (FE) analyses. The results are presented in a form which can be compared with the 'Ground Reaction Curve' (GRC). The results highlight the importance of the ratio of the embankment height over the centre-to-centre pile spacing. A better understanding of the variation for the performance of the piled embankment along with the change of the embankment height has been achieved. The analysis has been then extended to involve the single and multiple layers of reinforcement. The magnitude of the vertical load carried by the reinforcement and the tension mobilised in the reinforcement has been explored. This monograph adopts the British standard BS 8006 predictions of reinforcement tension derived from the 'Marston' and 'Hewlett and Randolph' approaches as a basis for embankment load exerted on the reinforcement for a wide range of piled embankment geometries. The predictions are compared with the results obtained from the three-dimensional finite element analyses. A suggested modification to the BS 8006 formula for predicting the reinforcement sag was hence proposed. The potentially beneficial bearing capacity of the subsoil beneath the embankment is also incorporated, both in the finite-element predictions and as a simple modification to the BS 8006 predictive method. The concept of an interaction diagram and the corresponding equation for equilibrium including arching, reinforcement, and subsoil which is used in the design is modified based on the findings.

Good design practice is necessary to prevent excessive piled embankment settlements. This monograph potentially aims to guide engineers engaged in designing and construction of the geotechnical structures located at the area containing soft soils where the piled supported and geosynthetic reinforced embankment may be used. This monograph can enhance the understanding of the load transfer mechanism in piled embankment. The mobilisation mechanism of soil arching and the contribution of the geosynthetic reinforcement and the subsoil to the bearing capacity of the embankment are also promoted. The research results provide a scientific basis and technical support for the settlement control of embankment for the design life of the infrastructure in soft soil areas.

Nanjing, China   Yan Zhuang
September 2024

Contents

1   **Introduction, Objectives**   1
    References   2

2   **Overview of the Piled Embankment Analysis Method**   5
  2.1   Introduction   5
  2.2   Arching Concept   6
      2.2.1   Rectangular Prism: Terzaghi (1943) and McKelvey
          (1994)   7
      2.2.2   Semicircular Arch: Hewlett and Randolph (1988)   10
      2.2.3   Positive Projecting Subsurface Conduits: BS8006
          (1995) and Marston's Equation   12
      2.2.4   Rectangular Pyramid Shaped Arching: Guido Method
          (1987)   14
      2.2.5   Other Mechanisms   16
  2.3   Introduction to the Ground Reaction Curve   19
      2.3.1   Initial Arching   22
      2.3.2   Break Point and Relative Arching Ratio   22
      2.3.3   Maximum Arching   23
      2.3.4   Loading Recovery Stage   23
      2.3.5   Ultimate State   23
  2.4   Reinforcement   24
      2.4.1   Introduction   24
      2.4.2   Methodology   25
      2.4.3   'Interaction Diagram'   27
  2.5   Summary   27
  References   29

3   **Ground Reaction Curve in Plane Strain and Three-Dimensional**
   Conditions   31
  3.1   Introduction   31
  3.2   Analyses Presented   32
      3.2.1   For Plane Strain Condition   32

         3.2.2   For Three-Dimensional Condition   34
   3.3   Results   37
         3.3.1   For Plane Strain Condition   37
         3.3.2   For Three-Dimensional Condition   46
   3.4   Summary   53
   References   54

4 **Geogrid Reinforced Piled Embankment in Plane Strain**
   and Three-Dimensional Conditions   57
   4.1   Introduction   57
   4.2   Calculation of Reinforcement Tension in BS8006 (2010)   59
         4.2.1   Embankment Arching   59
         4.2.2   Subsoil Support   60
         4.2.3   Determination of W_T   60
         4.2.4   Determination of T_{rp}   61
   4.3   Finite-Element Analyses   62
         4.3.1   For Plane Strain Condition   62
         4.3.2   For Three-Dimensional Condition   65
   4.4   Results   67
         4.4.1   For Plane Strain Condition   67
         4.4.2   For Three-Dimensional Condition   76
   4.5   Summary   88
   References   89

5 **Reinforced Piled Embankment with Subsoil in Plane Strain**
   and Three-Dimensional Conditions   91
   5.1   Introduction   91
   5.2   Calculation of Reinforcement Tension Including the Effect
         of Subsoil   92
   5.3   Analyses Presented   94
         5.3.1   For Plane Strain Condition   94
         5.3.2   For Three-Dimensional Condition   97
   5.4   Results   100
         5.4.1   For Plane Strain Condition   100
         5.4.2   For Three-Dimensional Condition   106
   5.5   Summary   112
   References   112

6 **Discussion of Results**   115
   6.1   Introduction   115
   6.2   Summary of Results   115
         6.2.1   Piled Embankment   115
         6.2.2   'Reinforced' Piled Embankment   116

6.2.3   'Reinforced' Piled Embankment with Subsoil   116
6.3   Comparison of General Trends of Behaviour as $h/(s - a)$ Varies ...   116
6.4   Comparison of the Value of $\sigma_s/\gamma(s - a)$ at the Point
        of Maximum Arching for Medium Height Embankments   117
6.5   Equation for Equilibrium Including Arching, Reinforcement
        and Subsoil   118
6.6   Case Studies   119
        6.6.1   Second Severn Crossing   119
        6.6.2   Construction of Apartments on a Site Bordering River
                    Erne, Northern Ireland   121
        6.6.3   A650 Bingley Relief Road   122
        6.6.4   A1/N1 Flurry Bog   123
        6.6.5   Case Study Comparison   123
6.7   Summary   128
References   129

7   Conclusions   131
    References   132

Notations

Dimensions

a The pile cap width (m)
h The height of embankment (m)
h_s The thickness of subsoil (m)
h_w The thickness of working platform (piling mat) (m)
l The length of the span (m)
s The centre-to-centre spacing of pile caps (m)

Vertical Stress

σ_a The stress at the base of the embankment due to the action of arching alone (i.e. from the Ground Reaction Curve) (kN/m^2)
σ_r The stress carried by geogrid (where this exists) (kN/m^2)
σ_s The vertical stress in the subsoil beneath the embankment (kN/m^2)
σ_u The vertical stress supporting the embankment (reinforced piled embankment with subsoil) (kN/m^2)
σ_w The stress acting on the subsoil due to the working platform (any imported material below the pile cap level, which is hence not affected by arching) (kN/m^2)
w_s The surcharge at the surface of embankment (kN/m^2)

Settlements

δ The compatible settlement in the interaction diagram (m)
δ_{ec} The settlement at the top of the embankment at the centreline above the pile cap (plane strain) (m)

δ_{ec} The settlement at the top of the embankment at the centre above the pile cap (3D) (m)

δ_{em} The settlement at the top of the embankment at the midpoint between piles (plane strain) (m)

δ_{em} The settlement at the top of the embankment at the midpoint of the diagonal between piles (3D) (m)

δ_i The interface friction angle between the embankment fill and geogrid (m)

δ_s The maximum settlement of the subsoil at the midpoint between piles (plane strain) (m)

δ_s The maximum settlement of the subsoil at the midpoint of the diagonal between piles (3D) (m)

Material Parameters

J The stiffness of the geogrid (kN/m)

E_0 The one-dimensional stiffness of the subsoil (kN/m^2)

E_s The Young's modulus of subsoil (MN/m^2)

γ The unit weight of the soil (kN/m^3)

v The Poisson's ratio

ϕ The angle of internal friction of the soil (degrees)

ψ The kinematic dilation angle (degrees)

Others

c The cohesion intercept (kN/m^2)

W_T The uniform stress acting on the geogrid (kN/m^2)

K The earth pressure coefficient (dimensionless parameter)

K_a The active earth pressure coefficient (dimensionless parameter)

K_p The passive earth pressure coefficient (dimensionless parameter)

T_{rp} The constant horizontal component of tension in the geogrid (kN/m 'into the page')

ε The average strain based on the total extension in the geogrid

h_c Critical height of arching

C_c The arching coefficient, which depends on h and a

Chapter 1
Introduction, Objectives

It is becoming necessary to construct projects on sites that may once have been considered unacceptable in terms of geotechnical issues. This is typified by the need to construct embankments over soft clay foundations. Embankments constructed on soft ground (e.g. for road or rail transport) in the UK and throughout the world frequently use piles or similar long slender foundations to transmit loads through the compressible soil to a stronger stratum beneath. However, the construction of embankments on soft soils has three potential problems:

- Low strength of the foundation soil significantly limits the allowable load (embankment height) that can be applied with adequate short-term stability;
- High deformability and low permeability cause large settlements to develop slowly with the dissipation of the pore water pressure (consolidation);
- Low bearing capacity of soft soils may induce shearing failure and collapse of foundation under self-weight of embankment and surcharge load, hence damaging to the upper building structures.

The foundations generally cover only a few percent of the total construction area, and economy dictates that they should be separated as widely as possible. However, normally provision of a structural 'raft' or similar at the base of the embankment to ensure that its weight is transferred to the piles would be too expensive and otherwise undesirable.

Rather it is normally assumed that natural 'arches' will form in the embankment over the soft soil between the foundations, and prevent differential settlement at the embankment surface. Polymer 'geogrids' which act in tension at the base of the embankment are also adopted to make increased pile spacing acceptable.

However, as evidenced by a number of recent conference and journal publications (e.g. Love & Milligan, 2003; Naughton et al., 2008; Ellis & Aslam, 2009), there is continuing debate in the European and international geotechnical communities regarding the suitability of a number of potential design methods for piled embankments. This is particularly the case for low embankments over very soft soil,

© The Author(s) 2025

Y. Zhuang, *Arching in Piled Embankments Including the Effects of Reinforcement and Subsoil*, https://doi.org/10.1007/978-981-96-1198-0_1

as evidenced by the failure of a 'load transfer platform' for a housing development in Enniskillen, Northern Ireland (the subject of a presentation at the UK Institution of Civil Engineers in September 2006).

The aim of the monograph is to investigate the principles underlying the behaviour of piled embankments, including the effects of tensile reinforcement and the underlying subsoil. This ultimately aims to give additional guidance to designers on issues such as distribution of load and differential settlement. The principle aims are as follows:

- To examine various aspects of arching behaviour in a piled embankment, both in plane strain and three-dimension conditions.
- To examine the effect of geometrical parameters, particularly pile spacing and systematic variation with embankment height.
- To establish the additional contribution of tensile reinforcement and subsoil.

This monograph is organized as follows. This chapter reports the introduction, aims and objectives of this monograph. Chapter 2 summarises existing theories and research related to piled embankments. Chapters 3–5 introduce a series of analyses and present the corresponding results for the piled embankment, the reinforcement piled embankment, and the reinforced piled embankment with subsoil, building logically in complexity. Chapter 6 summarises the key results and compares them with other recent research. A new method of analysis is also applied to some case studies of actual piled embankments. Chapter 7 gives final remarks and conclusions.

References

Ellis, E. A., & Aslam, R. (2009). Arching in piled embankments: Comparison of centrifuge tests and predictive methods—part 1 of 2. In *Ground engineering* (pp. 34–38).

Love, J., & Milligan, G. (2003). Design methods for basally reinforced pile-supported embenkments over soft ground. In *Ground Engineering* (39–43).

Naughton, P., Scotto, M., & Kempton, G. (2008). Piled embankments: Past experience and future perspectives. In *Proceedings of the 4th European geosynthetics conference*, Edinburgh, UK, Paper number 184.

Chapter 2
Overview of the Piled Embankment Analysis Method

2.1 Introduction

As mentioned in Chapter one, the construction of embankments over soft soils could induce the potential problems of long-term stability and large settlements. One of the most promising solutions to these problems is to use piled embankments (see Fig. 2.1). In many cases, this method appears to be the most practical, efficient (low long-term cost and short construction time) and an environmentally-friendly solution for construction on soft soil. The field applications are mainly highways, railways and construction of areas of fill for industrial or residential purposes.

Piles are installed through the soft subsoil and transfer load to a more competent stratum at greater depth. The majority of the load from the embankment is carried by the piles and thus there is relatively little load on the soft subsoil. By using a piled embankment, the construction can be undertaken in a single stage without having to wait for the soft clay to consolidate. Settlements and differential settlements are also significantly reduced when the technique is used successfully. Figure 2.1 also shows the notation for geometry and settlement used in this monograph.

- a is the width of the pile cap (m)
- s is the centre-to-centre spacing (m)
- h is the height of the embankment (m)
- δ_s is the settlement of the subsoil at the midpoint between piles (m)
- δ_{ec} is the settlement at the top of the embankment at the centreline above the pile cap (m)
- δ_{em} is the settlement at the top of the embankment at the midpoint between piles (m)

This notation may also be extended to three dimensions. For instance, for square pile caps on a square grid in plan the pile cap is a square with side length a. The inclusion of a single or multiple layers of geogrid or geotextile reinforcement at the base of embankment is also considered in this monograph.

© The Author(s) 2025
Y. Zhuang, *Arching in Piled Embankments Including the Effects of Reinforcement and Subsoil*, https://doi.org/10.1007/978-981-96-1198-0_2

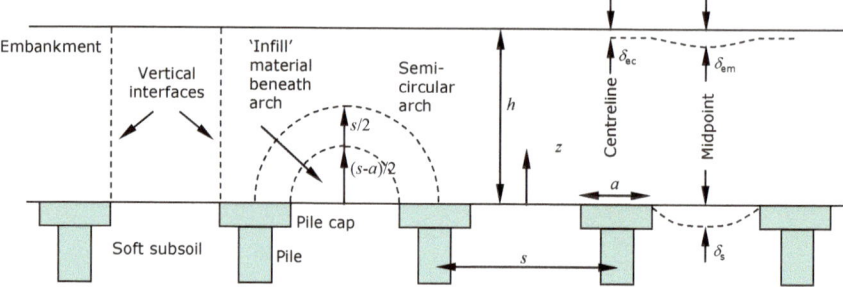

Fig. 2.1 Piled embankment showing potential arching mechanisms, and notation for geometry and settlement (δ) used in this monograph

2.2 Arching Concept

Differential settlement tends to occur between the relatively rigid piles and the soft foundation material. This causes the embankment fill material above the soft subsoil to settle more than the material above the piles. The differential settlement in the embankment fill will cause corresponding shear strain or shear planes so that vertical stress is redistributed from the embankment over the soft subsoil to the pile caps, hence reducing the load on the subsoil. The embankment is normally constructed from well-compacted granular material to maximise this arching effect.

A number of conceptual and analytical models of arching have been proposed, either in a general context or specifically for a piled embankment. As shown in Fig. 2.2, Terzgahi (1943) initially proposed vertical shear planes at either side of a 'trapdoor'. Hewlett and Randolph (1988) proposed a semicircular arch for piled embankments, whilst BS8006 (1995) is based on the analogy between the pile caps and a buried pipe.

Fig. 2.2 Stress state of a differential element (Terzaghi, 1943; Mckelvey, 1994)

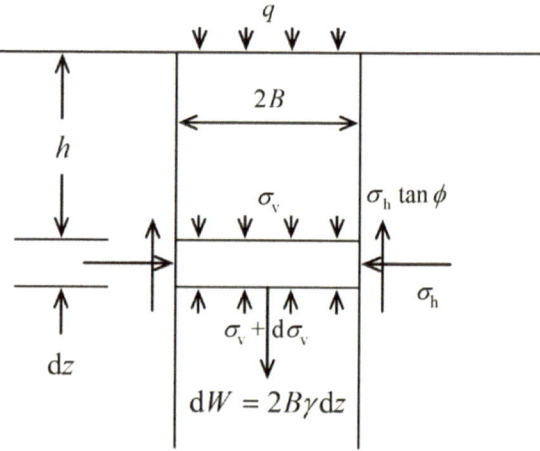

The concept of 'arching' of granular soil over an area where there is partial loss of support from an underlying stratum has long been recognised in the study of soil mechanics (e.g. Terzaghi, 1943). Its effect is widely observed, for instance in piled embankments. However, although this effect has been acknowledged for many decades, it remains quite poorly understood. There are a number of different models from different theoretical mechanisms and/or experimental data, but there does not yet exist a single method that can be agreed by the international geotechnical community. The following theoretical models of arching often consider plane strain conditions. The embankment fill is assumed to be a dry homogenous material. Thus, the total and effective stresses are equal.

2.2.1 Rectangular Prism: Terzaghi (1943) and McKelvey (1994)

Terzaghi was among the first theoreticians to define soil arching in his text "Theoretical Soil Mechanics" in 1943. Initially the vertical pressure at the base of the soil layer is everywhere equal to the nominal overburden stress. Terzaghi argued that gradually lowering a strip of support beneath the layer can cause yielding of the overlying material. The yielding material tends to settle, and this movement is opposed by shearing resistance along the boundaries between the moving and the stationary mass of sand. As a consequence the total pressure on the yielding strip is reduced whilst the load on the adjacent supports increases by the same amount (in terms of force).

When the strip has yielded sufficiently, a shear failure occurs along two surface of sliding (between the moving and stationary masses of sand) which rise from the outer boundaries of the strip potentially to the surface of the sand.

Terzaghi (1943) considers the equilibrium of a differential element and then integrates this through the depth (z) of the moving soil mass. See Fig. 2.2 where a rectangular soil element, having a thickness (dh) and weight (dw) is shown. The vertical stress applied to its upper surface is:

$$\sigma_v = \gamma h + q \tag{2.1}$$

where

σ_v the vertical stress (kN/m^2).
γ the unit weight of the soil (kN/m^3).
h the thickness of soil above the point (m).
q the surcharge acting at the surface of the soil (kN/m^2).

The corresponding normal stress on the vertical surface of sliding (σ_h) is given by:

$$\sigma_h = K\sigma_v \tag{2.2}$$

where

σ_h the horizontal stress (kN/m^2).
K the earth pressure coefficient (dimensionless parameter).

The shear strength of the soil is determined by (assuming the soil to be cohesionless):

$$\tau = \sigma_h \tan\phi \tag{2.3}$$

where

τ the shear strength (kN/m^2).
ϕ the angle of internal friction of the soil (degrees).

Resisting the movement of the soil element due to the applied stress and the weight of the element itself is the soil layer underlying this element ($\sigma_v + d\sigma_v$) and the shear strength of the soil adjacent to the element (τ) acting on both sides of the element. When the element is in equilibrium, the summation of the vertical forces must equal zero. Therefore, the vertical equilibrium can be expressed as:

$$\frac{d\sigma_v}{dz} = \gamma - K\sigma_v \frac{\tan\phi}{B} \tag{2.4}$$

where

$2B$ the width of the strip (m).
z the thickness of the soil overlying the element (m).

Using the boundary condition that $\sigma_v = q$ for $z = 0$, the partial differential equation can be solved as follows (Terzaghi, 1943 and later McKelvey, 1994):

$$\sigma_v = \frac{\gamma B}{K\tan\phi}\left(1 - e^{-K\tan\phi \, z/B}\right) + qe^{-K\tan\phi \, z/B} \tag{2.5}$$

If $q = 0$,

$$\sigma_v = \frac{\gamma B}{K\tan\phi}\left(1 - e^{-K\tan\phi \, z/B}\right) \tag{2.6}$$

The main problem with this method is that the coefficient of earth pressures K is not known and may vary through the depth of the sliding surface. Handy (1985) describes Terzaghi's approach as a 'lintel' rather than an arch. He also points out that there is a fundamental assumption behind Terzaghi's approach: that the vertical and horizontal stresses σ_v and σ_h equate to principal stresses σ_1 and σ_3. However, Krynine (1945) showed that the vertical and horizontal stresses could not be principal stresses if there is a plane of friction present. Krynine (1945) derived the following expression (Eq. (2.7)) for the earth pressure coefficient K.

$$K = \frac{1 - \sin^2\phi}{1 + \sin^2\phi} \tag{2.7}$$

Handy (1985) proposed that the shape of the arched soil is a catenary and suggested the use of the coefficient K_w instead of K, by considering an arch of minor principal stress. K_w is derived as:

$$K_w = 1.06 \left(\cos^2\theta + K_a\sin^2\theta\right) \tag{2.8}$$

where

$$\theta = 45° + \phi/2 \tag{2.9}$$

Russell et al. (2003) proposed that K could be conservatively taken as 0.5. More recently Potts and Zdravkovic (2008) proposed that $K = 1.0$ gave good correspondence with the results of plane strain finite element analyses of arching over a void. This does not seem to be consistent with frictional failure on a vertical plane. However, the assumption of failure on vertical planes is probably an oversimplification, particularly at the bottom of the soil layer near the void.

Figure 2.3a shows other work for an underground opening by McKelvey (1994) and Thigpen (1984). The width of the underground opening (a–b) is $2B$. The soil boundary (a–b) is assumed to have settlement Δ, and the remaining part is rigid. The base a–b is considered to be smooth so that $\tau = 0$ at $y = 0$.

The elasticity solution to this problem was obtained by Finn (Thigpen, 1984) by using the slip line method and considering a plane strain condition. The vertical stress compared to the nominal overburden stress γh from Finn's analyses is plotted in Fig. 2.3c. It is noted that the stress approaches infinity at the edge of the base from the elasticity solution of Finn. However, plastic flow would occur before this happened (Thigpen, 1984).

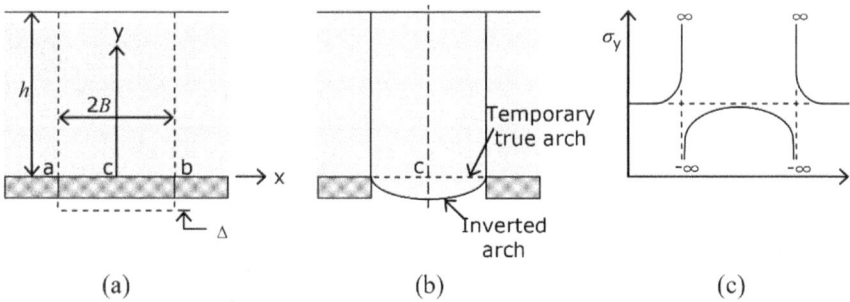

Fig. 2.3 Cross-section of a soil mass; **a** overlying the underground opening, **b** true soil arch collapses and the soil immediately above the void takes the shape of an inverted arch or catenary, **c** vertical stress distribution (**a, b** McKelvey, 1994, **c** Thigpen, 1984)

Thigpen (1984) described that as the base (a–b) in Fig. 2.3a yields, the compressive stress at the edge is steadily reduced (based on the elasticity solution of Finn, it changes to tensile). McKelvey (1994) proposed that momentarily just after the base yields, the soil remains its original position, forming a 'true arch', the soil directly over the underground opening is in tension. The soil tension arch can only last a finite period of time, which depends on the shear strength of the soil as well as other variables. McKelvey (1994) then states that the soil element in tension will ultimately fail as portions of the soil element begin to drop, leaving a small gap in the tension arch, ultimately forming the inverted arch as shown in Fig. 2.3b.

2.2.2 Semicircular Arch: Hewlett and Randolph (1988)

Hewlett and Randolph (1988) derived theoretical solutions based on observations from experimental tests of arching in a granular soil. Their analysis attempts to consider actual arches in the soil, as shown in Fig. 2.4 (rather than vertical boundaries as considered by Terzaghi). The 'arches of sand' transmit the majority of the embankment load onto the pile caps, with the subsoil carrying load predominantly from the 'infill' material below the arches. The arches are assumed to be semi-circular (in 2D) and of uniform thickness, with no overlap. The method also assumes that the pressure acing the subsoil is uniform.

The analysis considers the equilibrium of an element at the 'crown' of the soil arch (see Fig. 2.5a). Here the tangential (horizontal) direction is the direction of major principal stress and the radial (vertical) direction is the direction of minor principal stress, related by the passive earth pressure coefficient, K_p. Yielding is in the 'passive' condition since the horizontal stress is the major principle stress.

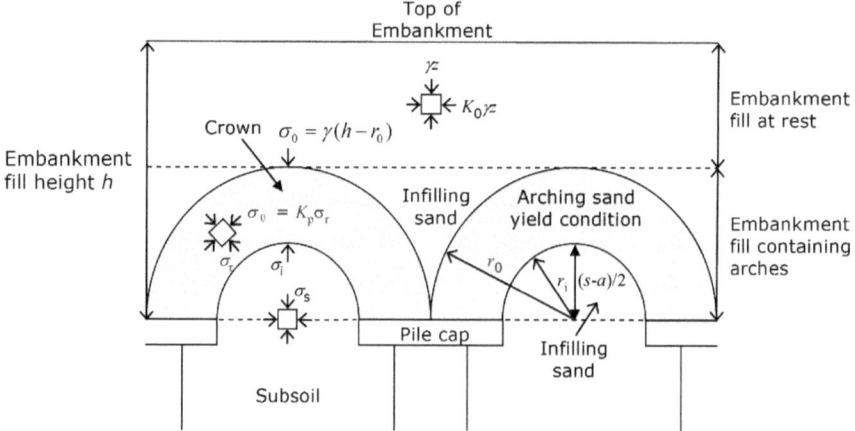

Fig. 2.4 Section through a piled embankment (Hewlett & Randolph, 1988)

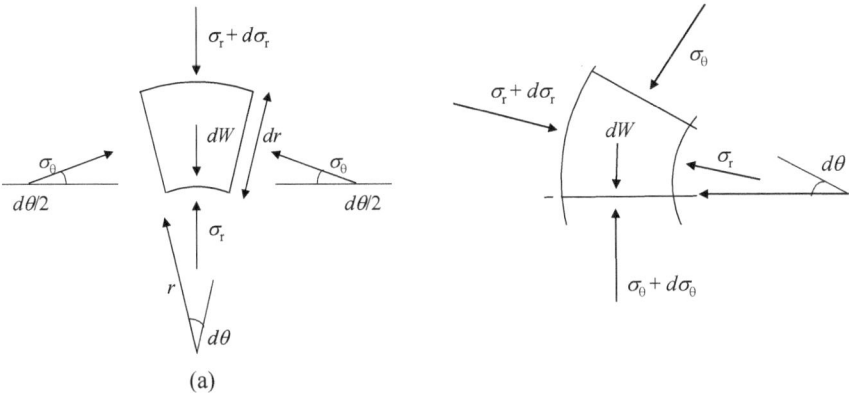

(a)

Fig. 2.5 Stresses on an element of soil arch; **a** an element of sand at the crown of an arch, **b** an element of sand above the pile cap

Considering the vertical equilibrium of this element, and using the boundary condition that the stress at the top of the arching layer is equal to the weight of material above acting on the outer radius of the arch gives a solution for the radial (vertical) stress acting immediately beneath the crown of the arch (σ_i). The vertical stress acting on the subsoil is then obtained by adding the stress due to the infilling material beneath the arch, based on the maximum height of infill $(s-a)/2$:

$$\sigma_s = \sigma_i + \gamma(s - a)/2 \tag{2.10}$$

The vertical stress (σ_s) is considered uniform here. In a refinement proposed by Low et al. (1994), a parameter is introduced to allow a possible non-uniform vertical stress on the soft ground.

At the pile cap (see Fig. 2.5b), the tangential (vertical) stress is the major principle stress, and the radial (horizontal) stress is the minor principle stress (the reverse of the situation at the crown). Again, equilibrium of an element of soil is considered, and in conjunction with overall vertical equilibrium of the embankment a value of σ_s is obtained in the limit when the ratio of the major and minor principle stresses is K_p. In fact yielding occurs in an 'active' condition, since the vertical stress is the major principle stress.

The initial solutions developed by Hewlett and Randolph (1988) are for a plane strain situation (see Fig. 2.4). However, equivalent 3-dimensional solutions for domes are also developed. It can be seen in Fig. 2.6 that different geometry is considered in the three-dimensional situation. For the crown of the arch, the maximum height of infill is now $(s - a)/\sqrt{2}$, thus the vertical stress acting on the subsoil (σ_s) is (see Fig. 2.6):

$$\sigma_s = \sigma_i + \gamma\,(s - a)/\sqrt{2} \tag{2.11}$$

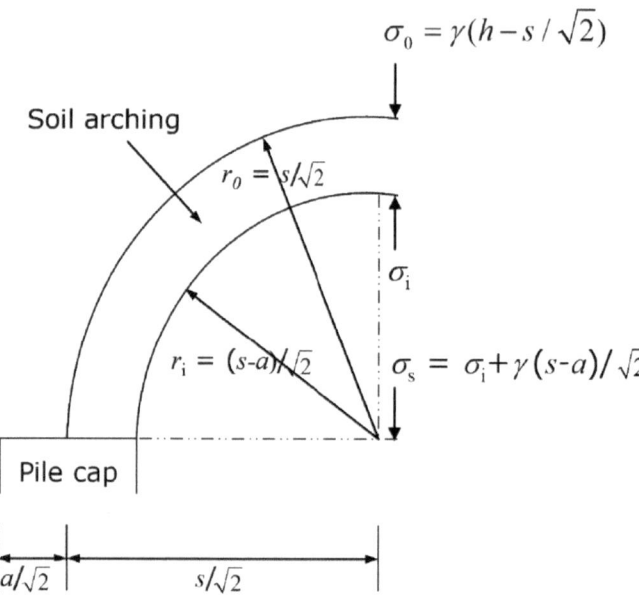

Fig. 2.6 Analysis of arching at the crown of a dome in a three-dimensional situation

Overall equilibrium of the embankment and 'active' yielding in the soil above the pile cap is used to obtain the value of σ_s in a similar manner to the plane strain analysis. This corresponds to the pile caps 'punching' into the underside of the embankment.

Hewlett and Randolph (1988) proved that for a 2-dimensional case, the critical point of the arch is always at the crown or at the pile cap. However, it is necessary to consider the value of σ_s resulting from failure of the arch either at the crown or pile caps—the largest value will be critical.

Hewlett and Randolph (1988) suggest that the pile spacing (s) should probably not exceed about 3 times the width of the pile caps (a), and the embankment fill should be chosen such that K_p is at least 3 (a friction angle of greater than 30°). In addition, in order to make optimum use of the piles, the spacing (s) should also be chosen such that the critical condition occurs at pile cap level, rather than at the crown of the arch. Also, the pile spacing (s) should not be greater than about half the embankment height (h) (Hewlett & Randolph, 1988).

2.2.3 Positive Projecting Subsurface Conduits: BS8006 (1995) and Marston's Equation

The method used in the British Standard for strengthened/reinforced soils and other fills (1995) to design geosynthetics over piles was initially developed by Jones et al.

(1990). A 2-dimensional geometry was assumed, which implies 'walls' in the soil rather than piles.

The British Standard differs from other methods by initially calculating the average stress on the pile cap itself rather than on the subsoil. BS8006 uses a modified form of Marston's equation for positive projecting subsurface conduits to obtain the ratio of the vertical stress acting on top of the pile caps to the average vertical stress at the base of the embankment ($\sigma_s = \gamma h$), using an equation normally used to calculate the reduced loads on buried pipes. The equation proposed by Marston was derived from field tests at the Engineering Experiment Station at Iowa State College in 1913.

For the 3-dimensional situation, and application to a piled embankment rather than a buried pipe, the result has been modified to give:

$$\frac{\sigma_c}{\gamma h} = \left(\frac{C_c a}{h}\right)^2 \qquad (2.12)$$

where

a the size (or diameter) of the pile cap (m).
C_c the arching coefficient, which depends on h and a.
h the height of the embankment (m).
γ the unit weight of the embankment fill (kN/m^3).
γh the nominal vertical stress at the base of the embankment (kN/m^2).
σ_c the vertical stress on the pile cap (kN/m^2).

It can be seen from Eq. (2.12) that the properties of the fill material have no effect on σ_c. It seems likely that the fill strength (which is accounted for in most methods) will have some impact, and it is likely that the result from BS8006 is only applicable to well compacted granular fill, with quite high frictional strength.

Vertical equilibrium requires that the combination of vertical stress on the pile caps (σ_c) and the subsoil (σ_s) must carry the embankment load. Thus, the overall vertical equilibrium is (for a 3D situation with pile caps with dimension a and spacing s):

$$\gamma h s^2 = \sigma_c a^2 + \sigma_s (s^2 - a^2) \qquad (2.13)$$

This can be re-arranged to give:

$$\sigma_s = \frac{\gamma h s^2 - \sigma_c a^2}{s^2 - a^2} \qquad (2.14)$$

Like many approaches BS8006 actually assumes that the 'subsoil' stress is carried by a geogrid at the base of the embankment. The distributed load W_T carried by the reinforcement between adjacent pile caps (see later Sect. 2.4) can be expressed as follows:

For $h > 1.4(s - a)$,

$$W_T = \frac{1.4 s \gamma (s - a)}{s^2 - a^2} \left[s^2 - a^2 \left(\frac{\sigma_c}{\gamma h} \right) \right] \qquad (2.15)$$

For $0.7(s - a) \le h \le 1.4(s - a)$,

$$W_T = \frac{s \gamma h}{s^2 - a^2} \left[s^2 - a^2 \left(\frac{\sigma_c}{\gamma h} \right) \right] \qquad (2.16)$$

These equations can be re-written as (substituting for σ_s from Eq. (2.14)):
For $h > 1.4(s - a)$,

$$W_T = \frac{1.4(s - a)}{h} s \left[\frac{\gamma h s^2 - \sigma_c a^2}{s^2 - a^2} \right] = \frac{1.4(s - a)}{h} s \, \sigma_s \qquad (2.17)$$

For $0.7(s - a) \le h \le 1.4(s - a)$,

$$W_T = s \left[\frac{\gamma h s^2 - \sigma_c a^2}{s^2 - a^2} \right] = s \, \sigma_s \qquad (2.18)$$

It has been proposed (e.g. Love & Milligan, 2003) that W_T can be calculated as $\sigma_s \, s$. These expressions are the same as this except that the first equation (for higher embankments) contains the factor $1.4(s - a)/h$. This effectively limits the height from the embankment considered to act on the subsoil to $1.4(s - a)$, instead of h. Thus, Love and Milligan (2003) concluded that the method does not satisfy vertical equilibrium, and also that it does not consider the condition at the crown of the arch.

In this method, a critical height $h_c = 1.4(s - a)$ is defined. If the embankment height is below the critical height, arching is not fully developed and all loads have to be supported by the geosynthetic membrane. Otherwise, it is assumed that all loads above the critical height are transferred directly to the piles as a result of arching in the embankment fill, and the soil weight below the critical height has to be supported by the geosynthetic membrane. This method does not allow $h_c < 0.7(s - a)$ to ensure that differential settlement does not occur at the top surface of the embankment.

2.2.4   Rectangular Pyramid Shaped Arching: Guido Method (1987)

This method is quite different from other methods of analysis for soil arching. The so-called 'Guido' design method is based on empirical evidence from model tests carried out with geogrid reinforced granular soil beneath a footing confined in a rigid box (see Fig. 2.7). The results suggest that multiple layers of geogrid reinforcement increase the bearing capacity, which could be interpreted as an improved angle of friction (or otherwise enhanced strength) for the composite soil/geogrid material

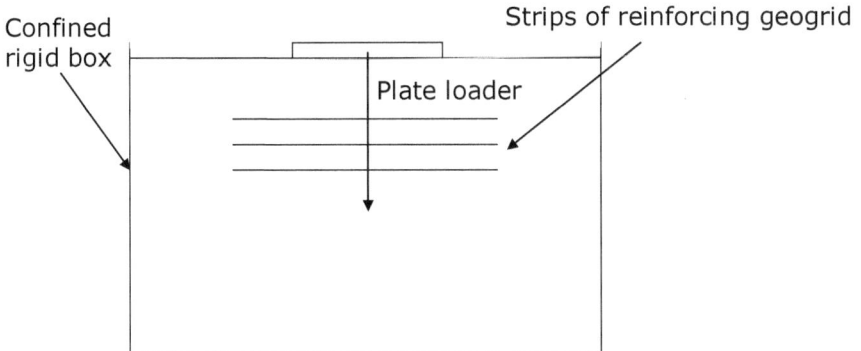

Fig. 2.7 Guido's experimental set-up (geogrid-reinforced sand in a confined, rigid box, the geogrid is used to improve the bearing capacity of the foundation soil)

(Slocombe & Bell, 1998). The 'load spread' angle in the reinforced soil beneath the footing was proposed to be 45° (Bell et al., 1994).

A piled embankment situation can be envisaged whereby the embankment soil is loading the pile caps, effectively inverting the arrangement above (Jenner et al., 1998). Thus, the load spread from the caps into the embankment is as shown in Fig. 2.8. The arch is a triangle with 45° angle in plane strain, and a similar pyramid in the three-dimensional case. Bell et al. (1994) applied this finding to evaluate an embankment with two layers of geosynthetic reinforcement supported on vibro-concrete columns (Stewart & Filz, 2005).

When this method has been employed in construction, numerous layers of relatively low strength geogrids at specific intervals are normally used, with significant compaction between each layer, so as to achieve the maximum lateral transmission of forces.

The stress on the subsoil (σ_s) results from the self-weight of the unsupported soil mass. The value is equal to the volume of the right-triangle/pyramid multiplied by the soil unit weight, and then divided by the area over which the soil prism acts. For the two-dimensional situation, the stress acting on the subsoil is:

$$\sigma_s = \frac{\gamma(s-a)}{4} \tag{2.19}$$

For the three-dimensional situation, the equation is modified to:

$$\sigma_s = \frac{\gamma(s-a)}{3\sqrt{2}} \tag{2.20}$$

It can be seen from Eqs. (2.19) and (2.20), that the height of the embankment has no effect on the pressure acing on subsoil. Additionally, the friction angle of the fill material (ϕ) is not considered in this case. The load spread angle above is assumed to be justified for compacted granular fill reinforced with multiple layers of geogrid.

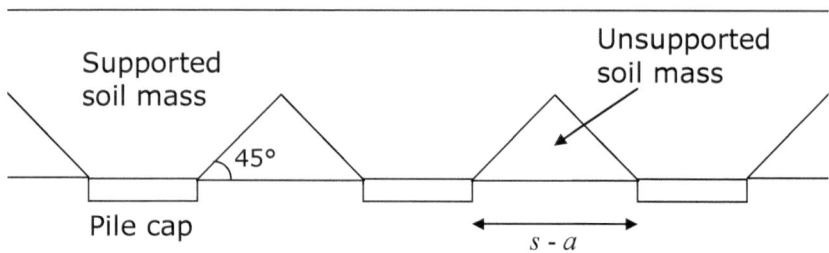

Fig. 2.8 The mechanism of load spreading from the pile caps through an embankment which is geogrid-reinforced near the base (shown in 2D)

The experiment was undertaken within a rigid box, and thus confining the granular material may have caused the material strength to be enhanced artificially.

This approach is similar to equations for arching at the crown of the embankment proposed by Hewlett and Randolph (1988) when an allowance for a thickness of infill material of $(s - a)/2$ (see Eq. (2.10)) and $(s - a)/\sqrt{2}$ (see Eq. (2.11)) was made in 2D and 3D cases respectively (based on the maximum thickness). However, in the Guido method, the average values of thickness are lower by factors of 2 and 3 respectively. Additionally, the Guido method does not consider any additional stress from the arch itself (σ_i, see Eq. (2.10) or Eq. (2.11)).

Love and Milligan (2003) point out that gravity in the embankment is operating in the opposite sense to that in Guido et al.'s laboratory tests; and the self-weight of the soil in the arch area therefore acts to reduce confinement. Additionally, the method requires the underlying subsoil and geogrid to have sufficient strength to completely carry the weight of the fill in the pyramid. Love and Milligan (2003) suggest that the Guido method may experience difficulties when dealing with situations where support from the exiting subsoil is very low or negligible. The Guido method concentrates more on reinforcement rather than on the actual physical arching process.

2.2.5 Other Mechanisms

Carlsson (1987) and Han and Gabr (2002) assume a trapezoidal shape (which is in effect a truncated triangle or pyramid). The Carlsson reference is presented in Swedish, but it is discussed by Rogbeck et al. (1998) and Horgan and Sarsby (2002) in English. In a plane strain situation, a wedge of soil is assumed under the arching soil, where the internal angle at the apex of the wedge is equal to 30° (see Fig. 2.9). The Carlsson Method adopts a critical height approach, and thus the additional overburden above the top of the wedge is transferred directly to the piles. As presented by Van Eekelen et al. (2003), the critical height was $1.87(s - a)$ in two-dimensions. Ellis

Fig. 2.9 Soil wedge
assumed by Carlsson (1987)
and Han and Gabr (2002)

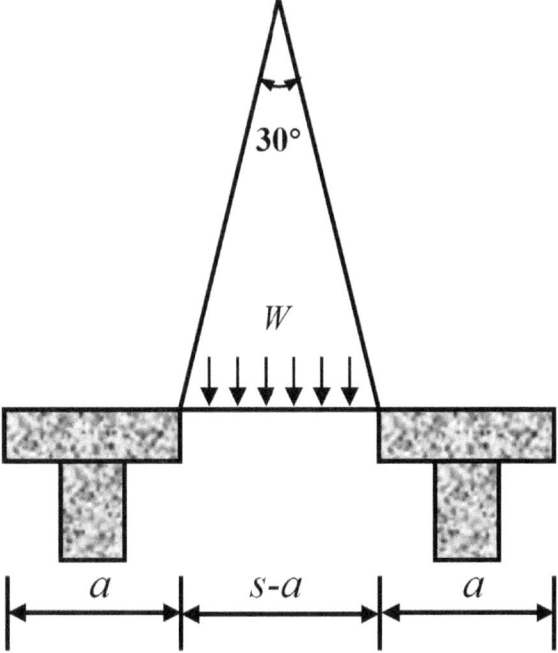

and Aslam (2009a) considered extending this theory to a 3-dimensional pyramid of
the same height, the average height would be $1.87/3(s - a) = 0.62(s - a)$, and hence
$\sigma_s/\gamma(s - a) = 0.62$.

Comparing with the Guido method (see Sect. 2.2.4), an angle of 45° to the horizontal is assumed for the edges of the pyramid. This is considerably lower than the
Carlsson method, and thus gives relatively low results, as shown in Eq. (2.20): $\sigma_s/$
$\gamma(s - a) = 0.24$. However, the Guido method does inherently assume that the soil is
reinforced.

The German standard (EGBEO, 2004) is based on a three-dimensional arching
model proposed by Kempfert et al. (1997), which appears similar to the Hewlett
and Randolph (1988) approach. However, the average vertical pressure acing on the
soft subsoil was obtained by considering the equilibrium of dome shaped arches of
varying size in the 'infill' material beneath a hemisphere (see Fig. 2.10). EGBEO
(2004) recommends the use of geosynthetic reinforcement but the arching effect and
the membrane tension are dissociated.

Naughton (2007) proposed a new method for calculating the magnitude of arching,
based on the 'critical height' for arching in the embankment. The critical height was
calculated assuming that the extent of yielding in the embankment fill was delimited
by a log spiral emanating from the edge of the pile caps (see Fig. 2.11). An expression
for the critical height (h_C) is then:

$$h_C = C(s - a) \tag{2.21}$$

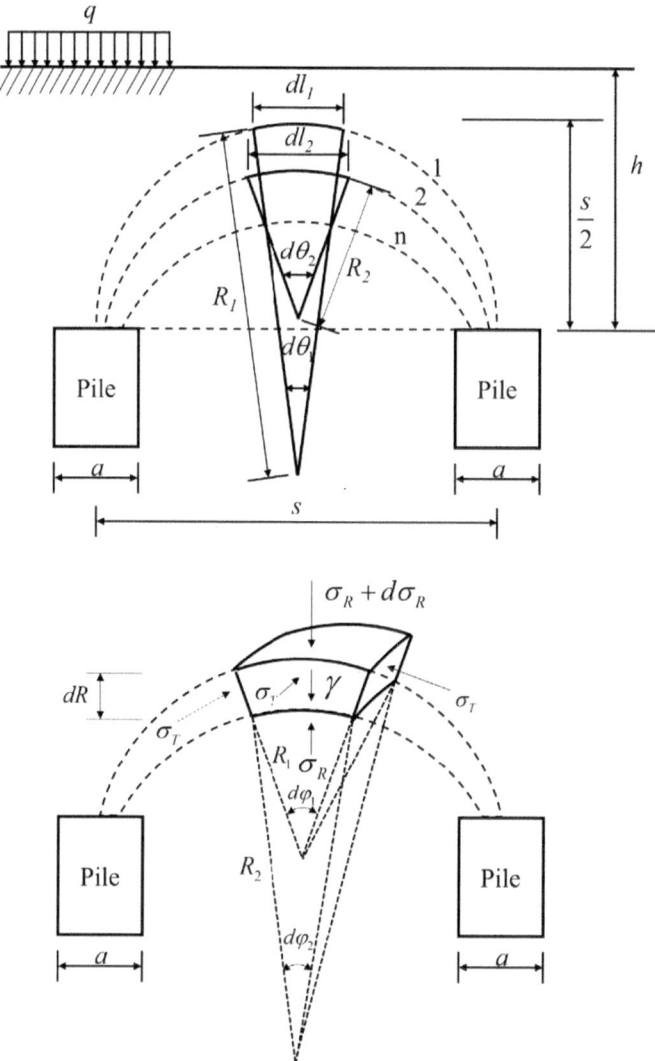

Fig. 2.10 Geometry of arching and equilibrium of stresses, German standard (EGBEO, 2004)

where

$$C = 0.5e^{\frac{\pi}{2}\tan\phi} \tag{2.22}$$

Fig. 2.11 Geometry of assumed log spiral shaped yield zone (Naughton, 2007)

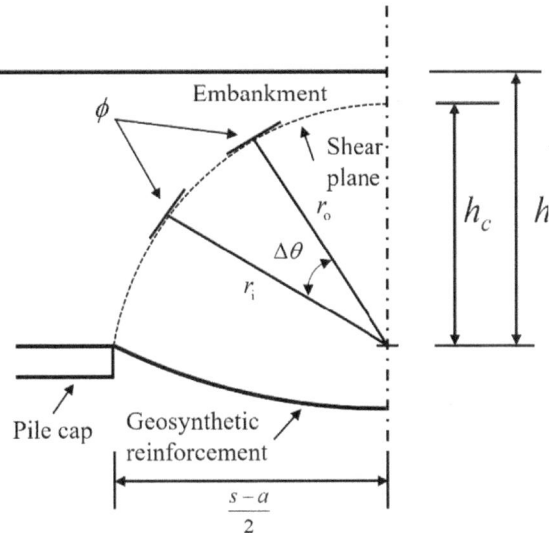

Naughton noted that the effect of ϕ on the critical height of the embankment was significant. Figure 2.12 shows the critical height varying from $1.24(s - a)$ to $2.40(s - a)$, as ϕ increases from $30°$ to $45°$. Naughton concluded that the critical height increases in proportion to the angle of friction.

Naughton suggests that the stress on the subsoil corresponds directly to the height of the zone of yielding so that

$$\sigma_s = \gamma h_C = \gamma C(s - a) \tag{2.23}$$

and hence

$$\frac{\sigma_s}{\gamma(s - a)} = C \tag{2.24}$$

However, this implies that the stress on the subsoil increases as the soil strength increases, which is not the expected trend of behaviour.

2.3 Introduction to the Ground Reaction Curve

The above methods consider that there is sufficient tendency for the soft subsoil to settle that arching of the embankment material will occur, but they do not specifically link arching with the amount of support from the subsoil. An interesting contrast to this is the concept of a 'Ground Reaction Curve' (GRC) used to determine the load on a plane strain underground structure such as a tunnel, Fig. 2.13.

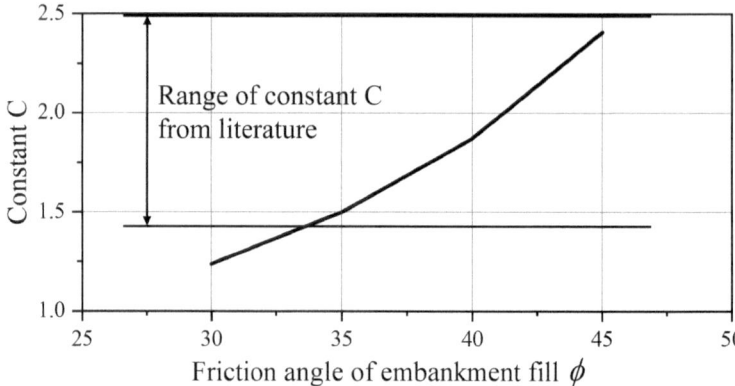

Fig. 2.12 Influence of ϕ on the critical height h_C of the embankment (Naughton, 2007)

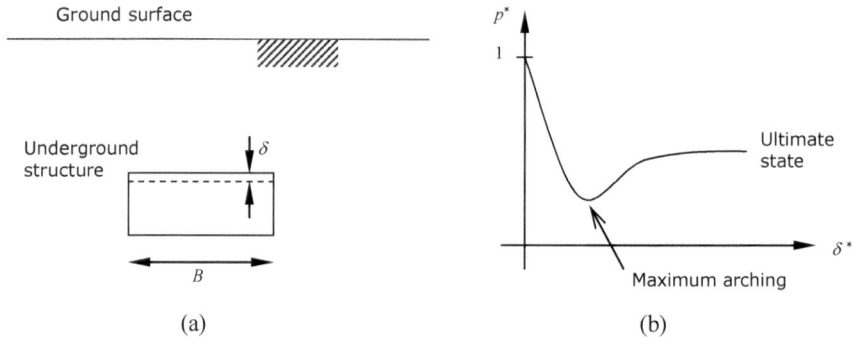

Fig. 2.13 Ground reaction curve for underground tunnel (Iglesia et al., 1999); **a** underground structure, **b** ground reaction curve (GRC)

By combining experimental data from centrifuge 'trapdoor' tests with some theories on load redistribution due to arching, a novel approach for determining the vertical loading on underground structures in granular soils has been developed (Iglesia et al., 1999). This approach creates the ground reaction curve, which is a plot of load on an underground structure as the structure deforms causing the soil above it to arch over it.

As can be seen in Fig. 2.14, it is proposed that as the trapdoor (or underground structure) is gradually lowered, the arch evolves from an initially curved shape (1) to a triangular one (2), before ultimately collapsing with the appearance of a prismatic sliding mass bounded by two vertical shear planes emanating from the sides of the trapdoor (3). Compared to analysis of a piled embankment the structure is analogous to the subsoil. It can be seen that the curved arch is similar to Hewlett and Randolph's semi-circular arch. The triangular arch is similar to Guido's triangular arch (although

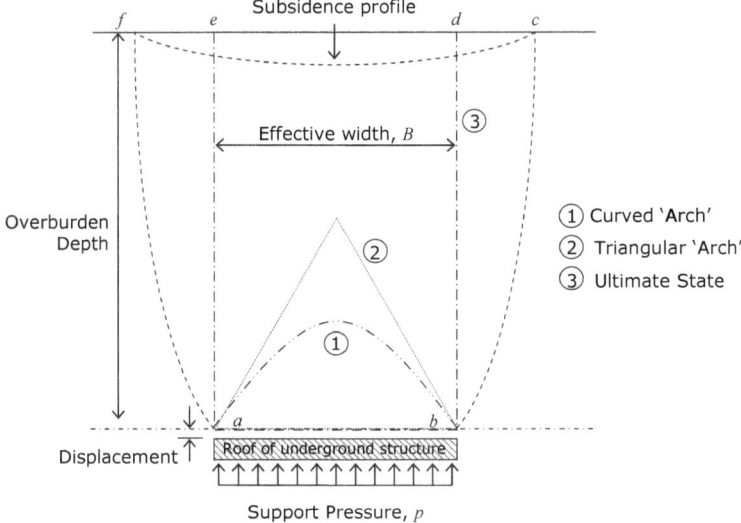

Fig. 2.14 Arching evolution (Iglesia et al., 1999)

the angle is somewhat greater than 45°), and the prismatic sliding mass is similar to Terzaghi's sliding block.

A methodology has been proposed by Iglesia et al. (1999) not only for determining the vertical loading on the structure, but also for relating this to the movement of the roof of the underground structure. This is referred to as a 'Ground Reaction Curve' (GRC) for the overlying soil. In the GRC, a dimensionless plot of normalised loading ($p*$) versus normalized displacement ($\delta*$) is used:

$$p^* = \frac{p}{p_0} \tag{2.25}$$

$$\delta^* = \frac{\delta}{B} \tag{2.26}$$

where

p the support pressure from the roof of the underground structure to the soil above (kN/m^2).

p_0 the nominal overburden total stress at the elevation of the roof derived from the thickness of overlying soil (and any surcharge at the ground surface) (kN/m^2).

B the width of the underground structure (m).

δ the settlement of the roof (m).

It can be seen in Fig. 2.15 that the GRC is divided into four parts—the initial arching phase, the maximum arching (minimum loading) condition, the loading recovery stage, and the ultimate state. These will be considered in turn below.

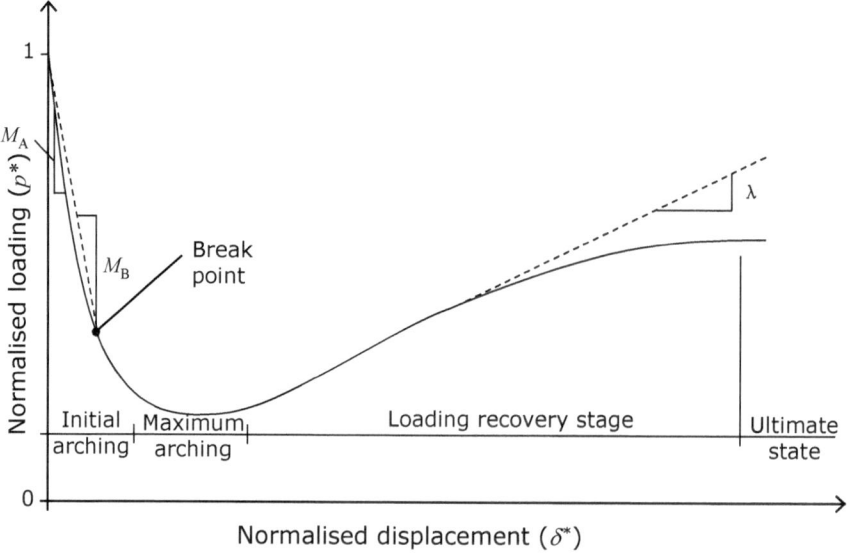

Fig. 2.15 Generalized Ground Reaction Curve (GRC) (Iglesia et al., 1999)

2.3.1 Initial Arching

As shown in Fig. 2.15, the GRC starts with the geostatic condition ($p_0 = \gamma h$). The initial 'convergence' of the soil toward the underground structure causes a fairly abrupt reduction in load on the structure. In this phase, the arch starts to form. A modulus of arching (M_A) is defined as the rate of initial stress decrease in the normalised plot. Iglesia et al. (1999) propose that based on the centrifuge trapdoor experiments with granular media, the modulus of arching has a value of about 125. Thus p^* tends to zero (or its minimum value, when this approaches zero) when $\delta^* > 1\%$.

2.3.2 Break Point and Relative Arching Ratio

As the underground opening converges toward a state of maximum arching (minimum loading), the GRC changes from the initial linear line to a curve (since p^* can only approach zero and certainly cannot be negative). Iglesia et al. (1999) propose a method of determining the approximate shape of this part of the curve—the reader is referred to the original paper for further details.

2.3.3 Maximum Arching

Maximum arching occurs when the vertical loading on the underground structure reaches a minimum. Iglesia et al. (1999) describe this corresponding to a condition in which a physical arch forms a parabolic shape just above the underground structure. In addition, this tends to occur when the relative displacement between the underground structure and the surrounding soil is about 2 to 6% of the effective width of the structure (B).

2.3.4 Loading Recovery Stage

This stage is the transition from the maximum arching (minimum loading) condition to the ultimate state (where the arch has become a prism with vertical stress sides as proposed by Terzaghi). Iglesia et al. (1999) characterise this stage by the load recovery index (λ). Based on centrifuge tests, they showed that the load recovery index increases with increasing B/D_{50} (D_{50} is the average particle size) and decreasing H/B. This aspect of behaviour is potentially of considerable significance, since it represents 'brittle' arching response.

2.3.5 Ultimate State

As the surrounding soil continually converges toward the underground structure, the arch will eventually collapse. Figure 2.14 shows the arching profile as presented by Finn's elasticity solution and Terzaghi. As the plane ab moves vertically the soil yields and the wedges aef and bdc move to the right and left respectively. As mentioned by Terzaghi, the real surfaces of sliding are curved and the real width of deformation at the surface of the soil layer may be considerably greater than the width of the yielding strip. Hence the surface of sliding must have a shape similar to that indicated in Fig. 2.14 by the lines af and bc. However, Terzaghi pragmatically assumed a sliding prism but maintained that it is on the 'unsafe' side (the friction along the vertical sections cannot be fully mobilised). Iglesia et al. (1999) use Eq. (2.6) (Terzaghi's method for the plane strain situation) to determine the ultimate stress on the structure.

2.4 Reinforcement

2.4.1 Introduction

In order to allow piles to be placed further apart, a reinforcing material can be included in the embankment fill between the piles. The vertical load carried by the reinforcement is transferred to the piles by tension as the reinforcement sags. The reinforcement can be 'geogrid' (with apertures) or 'geotextile'. The former term is used most widely in this monograph although generally any generic tensile reinforcement is implied.

A single layer of reinforcement may be used at or near the base of the embankment. Generally it is not placed directly on the pile caps due to the risk of damage. In this monograph it is assumed that a single layer of reinforcement is placed 100 mm above the pile cap.

Alternatively multiple layers of lower strength reinforcement may be distributed near the base of the embankment. This is often referred to as a 'Load Transfer Platform' (LTP). The premise is that this forms a zone of improved soil which enhances arching, particularly where geogrid which 'interacts' with the surrounding soil is used. Design of LTPs therefore often relies upon a contribution from the reinforcement beyond simple catenary action (sag). The geometry of LTP used in this monograph is shown in Fig. 2.16.

Whatever reinforcement is used, care is required that it is 'taut' during construction and filling so that tension will result immediately from subsequent sag. For LTPs careful compaction of the fill within the LTP is also sometimes considered to be of particular importance to further enhance interaction with the geogrid reinforcement.

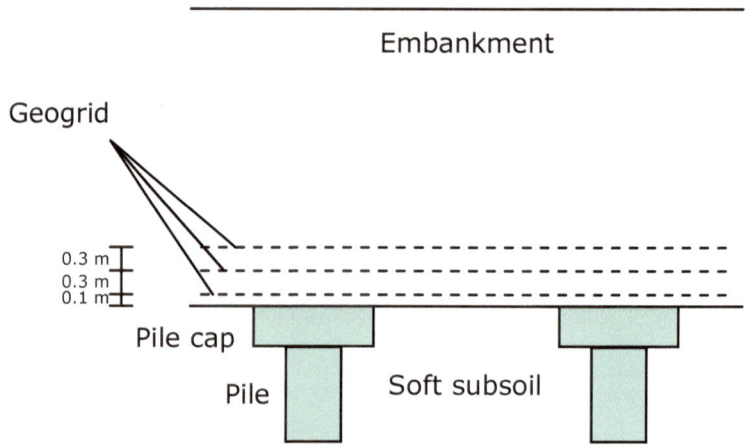

Fig. 2.16 Layout of geogrid in a piled embankment load transfer platform

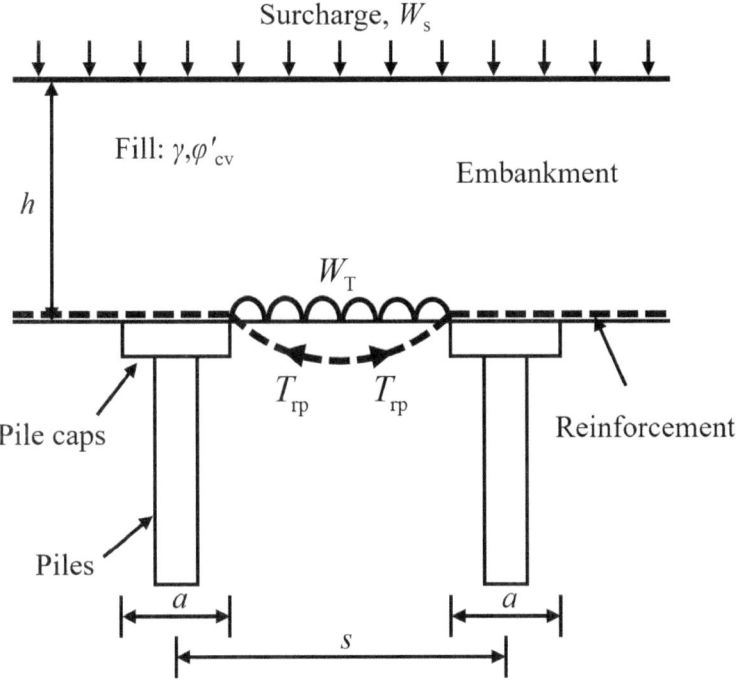

Fig. 2.17 W_T is the vertical load acting on a reinforcement strip between two adjacent pile caps (from BS8006)

Geogrid reinforcement is commonly used in soils. By placing geogrid at the base of the embankment, it is possible to improve support to the embankment. The tension will provide support between the pile caps (Fig. 2.17). At the edges of the embankment it also prevents lateral spreading (Hewlett & Randolph, 1988). However, these two functions are normally considered independently, and the former is of most interest in the context of this work.

2.4.2   Methodology

As described by Ellis and Aslam (2009b), the effect of additional capacity to carry vertical load from geogrid layer(s) could be added based on purely tensile response (but not accounting for any other interaction). They proposed that assuming the geogrid was subjected to a uniform vertical load and deforms as a parabola, using a plane strain approach the constant horizontal component of tension can be linked to the load acting on it as follows (e.g. Russell et al., 2003):

$$T_{rp} = \frac{W_T l^2}{8\delta_g} \tag{2.27}$$

where

T_{rp} the constant horizontal component of tension in the geogrid (kN/m 'into the page').
W_T the uniform stress acting on the geogrid (kN/m^2).
l the length of the span (m).
δ_g the maximum sag (vertical deflection) of the geogrid (m).

The average strain based on the total extension in the geogrid (ε) can be expressed in terms of the maximum sag as follows:

$$\varepsilon = \frac{8}{3}\left(\frac{\delta_g}{l}\right)^2 \tag{2.28}$$

Note that ε increases as the square of δ_g.
The equation links tension and strain in the geogrid assuming linear response:

$$T_{rp} = J\varepsilon \tag{2.29}$$

where

J the stiffness of the geogrid (kN/m).

Substituting for T_{rp} and ε from Eqs. (2.27) and (2.28) respectively

$$\frac{W_T l^2}{8\delta_g} = \frac{8J}{3}\left(\frac{\delta_g}{l}\right)^2 \tag{2.30}$$

This can be re-arranged to express how the load which can be carried theoretically increases with the sag:

$$W_T = \frac{64J}{3l}\left(\frac{\delta_g}{l}\right)^3 \tag{2.31}$$

where

W_T the assumed uniform stress acting on the geogrid (kN/m^2).
δ_g the maximum sag of the geogrid (m).
l the span of the geogrid (m).

2.4.3 'Interaction Diagram'

Before loading by the embankment, the soft foundation is in equilibrium, probably with hydrostatic pore water pressures. As embankment construction proceeds the soft subsoil will actually be virtually incompressible (it is likely to contain a significant fraction of clay and will therefore have low permeability). Thus, the vertical effective stress does not change and the increased stress is totally supported by an increase in pore water pressure. This excess pore water pressure causes water to flow out of the soil eventually, with accompanying settlement. In the absence of significant tendency for bearing failure the associated strain may be assumed to be one-dimensional.

Ellis and Aslam (2009b) introduced an interaction diagram, which combined the Ground Reaction Curve (GRC) with the effect of subsoil and/or geogrid or geotextile reinforcement by considering the normalised load on the subsoil and corresponding normalised settlement (see Fig. 2.18).

The settlement response of the subsoil to stress acting on it was determined from one-dimensional compression (the potential preconsolidation stress was also introduced; see Fig. 2.18a):

$$\sigma_s = E_0' \frac{\delta_s}{h_s} \tag{2.32}$$

where

h_s the thickness of the subsoil (m).
E_0' the one-dimensional stiffness of the subsoil (kN/m^2).
δ_s the settlement at the surface of the subsoil (m).

Figure 2.18a shows the combination of the GRC and the effect of the subsoil. Ellis and Aslam (2009b) argued that if the GRC and subsoil response meet, the subsoil is able to carry the 'remaining' embankment load accounting for arching at the given compatible settlement. Otherwise, there is not equilibrium.

Figure 2.18b shows the combination of GRC and the effect of subsoil and geogrid. Ellis and Aslam (2009b) concluded that if the stress from the embankment GRC can be carried by the combined response of the subsoil and geogrid (adding the stresses for each component at a given settlement), the lines intersect, then stability should result.

2.5 Summary

There are a number of theories to quantify arching in a piled embankment. Many authors have compared the methods for specific geometries and noted that they yielded differing results. However, they tend to focus on one or two specific geometries, or compare the results with numerical analyses, but without commenting systematically on the generic features of various methods.

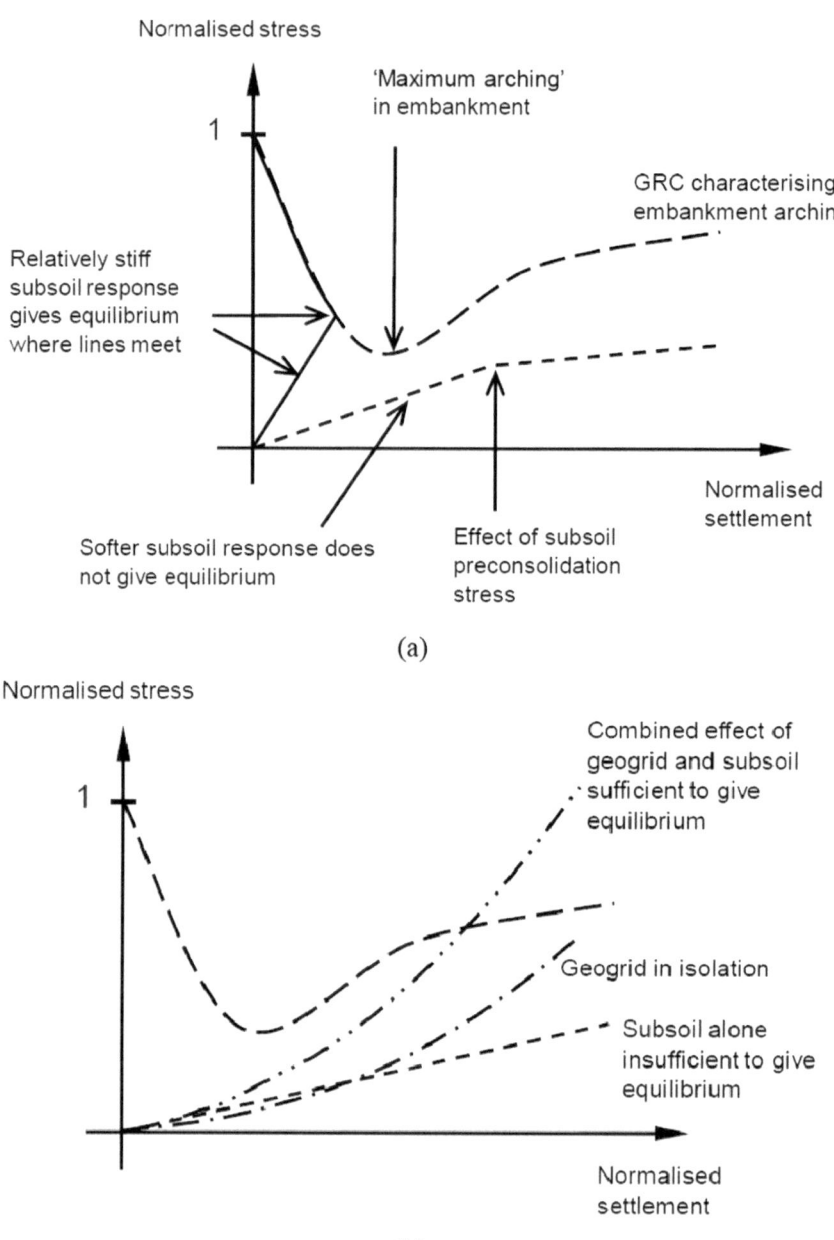

Fig. 2.18 Interaction diagrams for arching, subsoil and geogrid response (from Ellis & Aslam, 2009b); **a** schematic illustration of embankment and subsoil response to give equilibrium, **b** schematic illustration of combination of subsoil and geogrid response to potentially give equilibrium

Any effect due to the subsoil is generally neglected. However, this influence could be a major effect on the overall embankment response, which will be discussed in detail in the following chapters.

References

Bell, A. L., Jenner, C., Maddison, J. D., & Vignoles, J. (1994). Embankment support using geogrids with vibro concrete columns. In *Fifth international conference on geotextiles, geomembranes and related products* (pp. 335–338).

BS8006. (1995). *Code of practice for strengthened/reinforced soils and other fills.* British Standards Institution.

Carlsson, B. (1987). *Reinforced soil, principles for calculation.* Terratema AB, Linköping (in Swedish).

EGBEO (2004). Bewehrte ErdkÖrper auf punkt-und linienfÖrmigen Traggliedern, Entwurf Kapitel 6.9, 05/16/2004 version, nonpublished.

Ellis, E. A., & Aslam, R. (2009a). Arching in piled embankments: Comparison of centrifuge tests and predictive methods—Part 1 of 2. In *Ground engineering* (pp. 34–38), June 2009.

Ellis, E. A., & Aslam, R. (2009b). Arching in piled embankments: Comparison of centrifuge tests and predictive methods—Part 2 of 2. In *Ground engineering* (pp. 28–31), July 2009.

Guido, V. A., Knueppel, J. D., & Sweeny, M. A. (1987). Plate loading tests on geogrid—Reinforced earth slabs. In *Proceedings geosynthetics 87 conference*, New Orleans, pp. 216–225.

Han, J., & Gabr, M. A. (2002). Numerical analysis of geosynthetic-reinforced and pile-supported earth platforms over soft soil. *Journal of Geotechnical and Geoenvironmental Engineering*, 44–53.

Handy, R. L. (1985). The arch in soil arching. *ASCE Journal of Geotechnical Engineering, 111*(3), 302–318.

Hewlett, W. J., & Randolph, M. F. (1988). Analysis of piled embankments. In *Ground engineering* (pp. 12–18), April 1988.

Horgan, G. J., & Sarsby, R. W. (2002). The arching effect of soils over voids and piles incorporating geosynthetic reinforcement. In Delmas, Gourc, & Girard (Eds.), *Geosynthetics—7th ICG*, Swets and Zeitlinger, Lisse ISBN 90-5809-523-1, pp. 373–378.

Iglesia, G. R., Einstein, H. H., & Whitman, R. V. (1999). Determination of vertical loading on underground structures based on an arching evolution concept. In *Proceedings 3rd national conference on geo-engineering for underground facilities*, pp. 495–506.

Jenner, C. G., Austin, R. A., & Buckland, D. (1998). Embankment support over piles using geogrids. In *Proceedings 6th international conference on geosynthetics*, Atlanta, USA, pp. 763–766.

Jones, C. J. F. P., Lawson, C. R., & Ayres, D. J. (1990). Geotextile reinforced piled embankments. In D. Hoedt (Ed.), *Geotextiles, geomembranes and related products*, Balkema, Rotterdam, pp. 157–159.

Kempfert, H.G., Stadel, M., & Zaeske, D. (1997). Design of geosynthetic–reinforced bearing layers over piles. *Bautechnik* 74, Heft 12, pp. 818–825.

Krynine, D. P. (1945). Discussion of "stability and stiffness of cellular cofferdams", by Karl Terzaghi. *Transactions, ASCE, 110*, 1175–1178.

Love, J., & Milligan, G. (2003). Design methods for basally reinforced pile-supported embankments over soft ground. In *Ground engineering* (pp. 39–43), March 2003.

Low, B. K., Tang, S. K., & Choa, V. (1994). Arching in piled embankments. *ASCE Journal of Geotechnical Engineering, 120*(11), 1917–1938.

McKelvey, J. A. (1994). The anatomy of soil arching. *Geotextiles and Geomembranes, 13*, 317–329.

Naughton, P. J. (2007). The significance of critical height in the design of piled embankments. In *Proceedings of Geo-Denver 2007*, Denver, Colorado.

Naughton, P., Scotto, M., & Kempton, G. (2008). Piled embankments: Past experience and future perspectives. In *Proceedings of the 4th European geosynthetics conference*, Edinburgh, UK, September 2008, Paper number 184.

Potts, V. J., & Zdravkovic, L. (2008). Finite element analysis of arching behaviour in soils. In *The 12th international conference of international association for computer methods and advances in geomechanics (IACMAG)*, Goa, India, October 2008, pp. 3642–3649.

Rogbeck, Y., Gustavsson, S., Södergren, I., & Lindquist, D. (1998). Reinforced piled embankments in Sweden-design aspects. In *Proceedings of the 6th international conference on geosynthetics*, pp. 755–762.

Russell, D., Naughton, P., & Kempton, G. (2003). A new design procedure for piled embankments. In *Proceedings of the 56th Canadian geotechnical conference and the NAGS conference*, Winnipeg, MB, pp. 858–865.

Slocombe, B. C., & Bell, A. L. (1998). Setting on a dispute: Discussion on Russell and Pierpoint paper. In *Ground engineering* (pp. 34–36), March 1998.

Stewart, M. E., & Filz, G. M. (2005). Influence of clay compressibility on geosynthetic loads in bridging layers for column—Supported embankments. In *Proceedings of GeoFrontiers 2005*. ASCE.

Terzaghi, K. (1943). *Theoretical soil mechanics*. Wiley.

Thigpen, L. (1984). *On the mechanics of strata collapse above underground openings*. Lawrence Livermore National Laboratory, U.S. Department of Energy.

Van Eekelen, S. J. M., Bezuijen, A., & Oung, O. (2003). Arching in piled embankments; experiments and design calculations. In *BGA international conference foundations: Innovations, observations, design and practice*, Dundee, pp. 889–894.

Chapter 3
Ground Reaction Curve in Plane Strain and Three-Dimensional Conditions

3.1 Introduction

The concept of 'arching' in granular soil over an area where there is partial loss of support from an underlying stratum has long been recognised in the study of soil mechanics, but notably there is not one generally accepted approach to account for this phenomenon in the design of piled embankments. Piled embankments rely upon arching of the embankment material onto underlying piles, thus potentially significantly reducing load on the soft subsoil that more generally prevails beneath the embankment.

Figure 2.1 illustrates some basic concepts (represented in side elevation). Popular mechanisms of arching in the embankment include vertical interfaces (Terzaghi, 1943) or a semi-circular arch (Hewlett & Randolph, 1988). The latter becomes a hemisphere in three dimensions. As shown, there may be differential settlement at the surface of the embankment.

A number of authors (e.g. Naughton & Kempton, 2005; Stewart & Filz, 2005) have concluded that various analytical methods proposed for arching in piled embankments yield different results in specific situations, but the discussions on systematic differences remains absent. A recent review which attempts to rationalise the generic outcome of various methods is given by Ellis and Aslam (2009a). Naughton et al. (2008) considered the historical development of analysis of piled embankments. It was concluded that the design of piled embankments was complex and not yet fully understood, particularly when considering the three-dimensional effects and the support from the subsoil.

Finite-element modelling of a piled embankment in plane strain and in three-dimensional were reported in this chapter to investigate the arching in this system. The 'subsoil' was not explicitly modelled, but the support of the subsoil was simulated by a vertical stress acting on the underside of the embankment. The results are presented in a form which can be compared with the 'Ground Reaction Curve' (GRC) proposed by Iglesia et al. (1999), Sect. 2.3 (Fig. 2.15). This approach was originally proposed

© The Author(s) 2025   31
Y. Zhuang, *Arching in Piled Embankments Including the Effects of Reinforcement and Subsoil*, https://doi.org/10.1007/978-981-96-1198-0_3

to consider arching over an underground structure. However, settlement of the subsoil beneath a piled embankment can be compared to deformation of the underground structure, since both tend to cause arching of the material above.

3.2 Analyses Presented

The finite element analysis models of piled embankment were established in the following section, in which the plane strain condition and three-dimensional condition were both considered and analysed. The model sizes, material parameters for embankment, finite element mesh and boundary conditions, together with numerical analyses cases were all reported in detail.

3.2.1 For Plane Strain Condition

The analyses demonstrated here were undertaken in plane strain using Abaqus Version 6.6. Figure 3.1 shows a typical mesh for the embankment, with height $h = 5.0$ m and pile spacing $s = 2.5$ m. There are 1474 eight noded, reduced-integration, two-dimensional, quadratic solid elements (CPE8R). As for reduced-integration elements, Abaqus evaluates the material response at each integration point in each element (Abaqus Analysis User's Manual, Version 6.6). Reduced-integration elements were chosen both for computational efficiency, and because second-order reduced-integration elements generally yield more accurate results than the corresponding fully integrated elements.

The vertical boundaries represent lines of symmetry at the centreline of a support (pile cap), and the midpoint between supports (see Sect. 2.1, Fig. 2.1). Hence there is a restraint on horizontal (but not vertical) movement at these boundaries. No boundary conditions are imposed at the top (embankment) surface, and no surcharge is considered to act here.

The bottom boundary represents the base of the embankment, which is underlain by a half pile cap (width $a/2$) on the left, and subsoil (width $(s - a)/2$) at the right. The pile cap is assumed to provide rigid restraint to the embankment, both horizontally and vertically. The assumed uniform vertical stress in the subsoil beneath the embankment (σ_s) is used to control the analysis—the subsoil itself was not actually modelled in this Chapter.

The pile cap width (a) was fixed at 1.0 m and the centre-to-centre spacing (s) was 2.0, 2.5 or 3.5 m. The embankment height (h) was varied using values in the range 1.0–10 m. Throughout the analyses minimum and maximum element sizes were approximately 0.002 and 0.012 m^3/m respectively. This corresponds to side lengths in the plane strain section of approximately 50–150 mm.

The embankment material was assumed to be granular (and hence with predominantly frictional strength), and was modelled using the linear elastic and Mohr

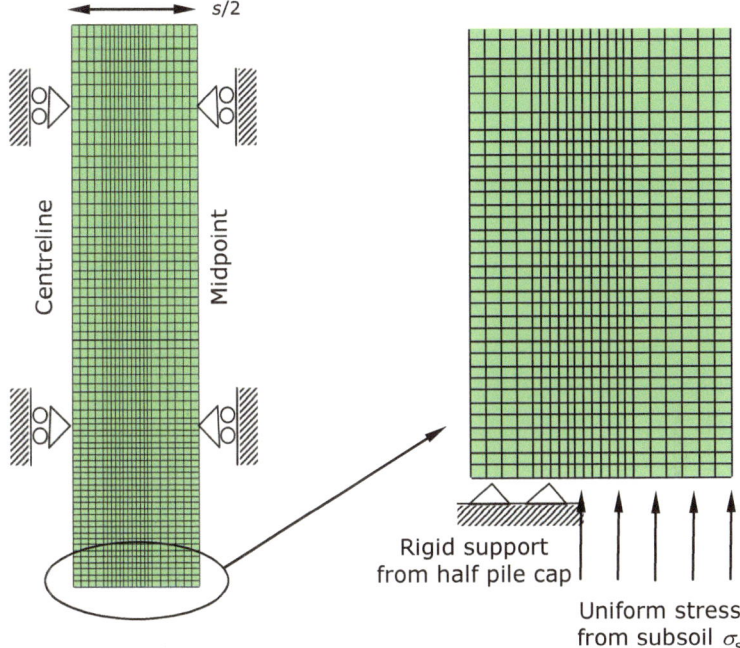

Fig. 3.1 Typical finite element mesh ($h = 5$ m, $s = 2.5$ m) and boundary conditions

Coulomb (c', ϕ') parameters shown in Table 3.1. For $s = 2.5$ m the effect of increasing ϕ' to 40°, or increasing the kinematic dilation angle at yield (ψ) to 22° was also considered. This value of dilation angle is quite high, and did not reduce with ongoing deformation at yield (compared to actual soil where dilation is a transient effect). However, the aim of this analysis was to assess the sensitivity to dilation rather than to model the effect accurately, which would require a sophisticated constitutive model using complex and probably uncertain input parameters. The granular material was assumed to be dry and hence pore water pressures were not considered.

The sequence of analysis was straightforward. First the in-situ stresses were specified (based on a unit weight of 17 kN/m³ and a K_0 value of 0.5) using the 'Geostatic' command in Abaqus. The K_0 value is based on a nominal 'at rest' value taken as $(1-\sin\phi')$. Initially σ_s was specified as the nominal vertical stress at the base of the embankment to give equilibrium with the in-situ stresses. This value was then reduced

Table 3.1 Material parameters for embankment fill

Unit weight (kN/m³)	Initial earth pressure coefficient	Young's modulus (MN/m²)	Poisson's ratio	Cohesion intercept (kN/m²)	Friction angle (°)	Kinematic dilation angle at yield (°)
17.0	0.5	25	0.20	1	30	0 (or 22)

Table 3.2 Summary of analyses reported in plane strain

h(m)	$s = 2$ m	$s = 2.5$ m	$s = 3.5$ m	$s = 2.5$ m	$s = 2.5$ m
	$c' = 1$ kN/m^2 $\phi' = 30°$ $\psi = 0°$	$c' = 1$ kN/m^2 $\phi' = 30°$ $\psi = 0°$	$c' = 1$ kN/m^2 $\phi' = 30°$ $\psi = 0°$	$c' = 1$ kN/m^2 $\phi' = 40°$ $\psi = 0°$	$c' = 1$ kN/m^2 $\phi' = 30°$ $\psi = 22°$
Sub-plot	(a)	(b)	(c)	(d)	(e)
1.0		✓			
1.5	✓	✓	✓		
2.0		✓		✓	✓
2.5	✓	✓	✓		
3.5	✓	✓		✓	✓
5.0		✓	✓	✓	✓
6.5		✓	✓		
8.0		✓			
10.0	✓	✓	✓	✓	✓

(generally allowing Abaqus to determine increment size automatically) to mimic loss of support from the subsoil. The subsoil in question is generally of low permeability, and thus this process has direct analogy with consolidation of the subsoil, which causes arching of the embankment material onto the pile caps. All analyses presented in plane strain are summarised in Table 3.2. Variations to the 'standard' parameters are highlighted in bold.

3.2.2   For Three-Dimensional Condition

A three-dimensional model of a piled embankment is also reported in this Chapter. This is an extension of the plane strain analyses, where the 'Ground Reaction Curve' (GRC) for arching in the embankment was studied without consideration of reinforcement or subsoil.

There is less support from the pile caps in a three-dimensional situation compared to the plane strain condition. Figure 3.2 shows the three-dimensional geometry in plan. A 'unit cell' with square dimensions of half the centre-to-centre pile spacing ($s/2$) was used in the analyses, and the pile caps are assumed to be square with width 'a', and thus the total pile cap area per unit is a^2 and the remaining subsoil area is ($s^2 - a^2$). Hence the mesh representing the embankment was a simple cuboid with height, h (the embankment thickness).

As shown in Fig. 3.2 the analysis uses lines of symmetry to consider a model which is one-quarter of this unit (and one-quarter of a pile cap). In fact, it would also be possible to bisect this model with a 45° line, halving the computation effort.

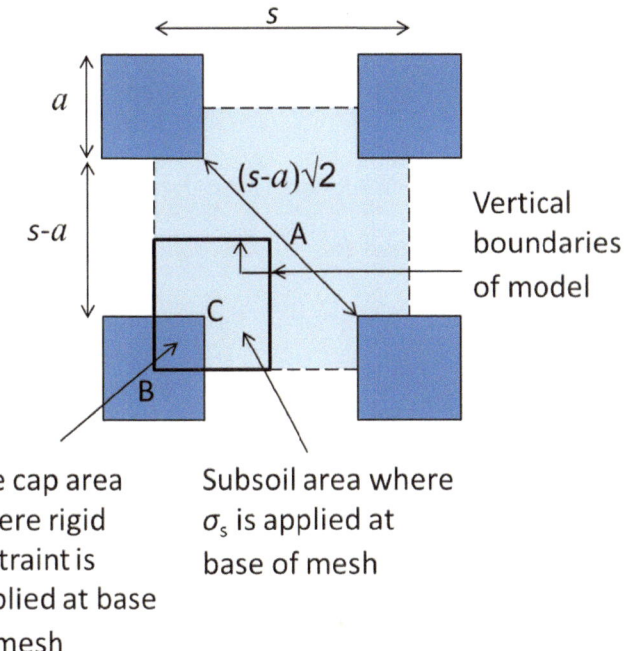

Fig. 3.2 Plan view of square pile cap layout, showing vertical boundaries of finite element model

However, owing to the corresponding complications in mesh generation this was not done.

Figure 3.3 shows the mesh for the embankment with $h = 3.5$ m and $s = 2.5$ m. There are 5040 twenty-noded, reduced-integration, three-dimensional, quadratic brick solid elements (C3D20R) with dimension of the order 100 mm. Halving this dimension had a minimal effect on results (e.g. increase of 0.03 in ultimate normalised stress in Fig. 3.8b), indicating that it was sufficiently small.

Corresponding to conditions of symmetry (Fig. 3.2), the vertical boundaries of the cuboid were restrained against horizontal movement normal to each face. As assumed in the plane strain condition, the top of the embankment surface (face 1) can move freely in all directions, and there is no surcharge acting here. The base of the mesh (Face 4 and face 5) consists of one-quarter of a pile cap, and the interface with the subsoil (Fig. 3.2). The pile cap was modelled as a rigid restraint to the overlying embankment in the vertical and horizontal directions. Over the area of the subsoil there was no kinematic restraint to the underside of the embankment, but a uniform vertical stress (σ_s) was applied acting upwards to support the fill and represent the presence of the subsoil. This 'subsoil stress' was used to control the analysis, reducing from an initial value equal to the nominal vertical stress from the embankment ($\sigma_s = \gamma h$). As expected, non-uniform settlement of the embankment

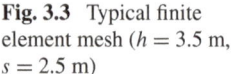

Fig. 3.3 Typical finite element mesh ($h = 3.5$ m, $s = 2.5$ m)

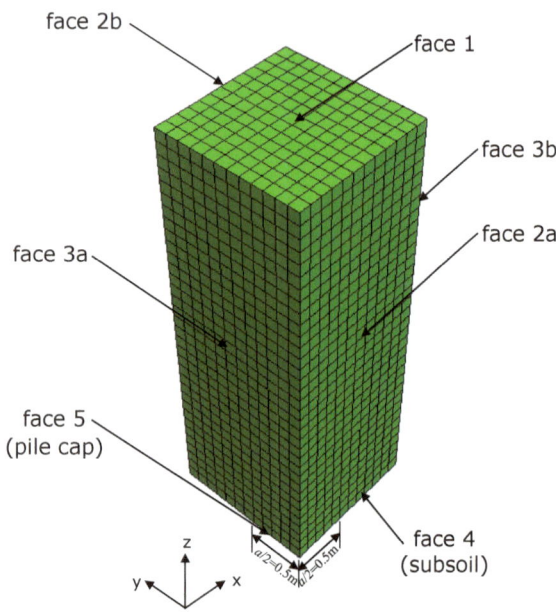

was observed at the base of the mesh (embankment), and was largest at the corner—point A (Fig. 3.2). All analyses considered in three dimensional are summarised in Table 3.3.

The sequence of analysis was the same as the plane strain case. First the in-situ stresses were specified (again based on a unit weight of 17 kN/m³ and K_0 value of 0.5). Initially σ_s was specified as the nominal vertical stress at the base of the embankment to give equilibrium with the in-situ stresses, but this value was then

Table 3.3 Summary of analyses in three-dimensional condition

h (m)	$s = 2.0$ (m) $c' = 1$ kN/m² $\phi' = 30°, \psi = 0°$	$s = 2.5$ (m) $c' = 1$ kN/m² $\phi' = 30°, \psi = 0°$	$s = 3.5$ (m) $c' = 1$kN/m² $\phi' = 30°, \psi = 0°$
Subplot	(a)	(b)	(c)
1.0		✓	
1.5	✓		
2.0		✓	
2.5	✓		✓
3.5	✓	✓	
5.0		✓	✓
6.5		✓	✓
8.0			
10.0	✓	✓	✓

reduced (allowing Abaqus to determine increment size automatically) to mimic loss of support from the subsoil.

3.3 Results

The numerical simulation results including Ground Reaction Curves, midpoint profile of earth pressure coefficient, ultimate stress on the subsoil, and settlement at the subsoil and surface of the embankment for the plane strain and three-dimensional conditions will be presented and discussed in the following section. The effect of embankment height, pile spacing on the arching was analysed, and the vertical extent of arching in the embankment will be determined. Furthermore, the method to estimate the stress on the subsoil will be proposed in the following section.

3.3.1 For Plane Strain Condition

3.3.1.1 Ground Reaction Curves

Figure 3.4 shows normalised ground reaction curves (GRC) broadly equivalent to the approach described in Iglesia et al. (1999). The subsoil stress (σ_s) is normalised by the nominal overburden stress at the base of the embankment (γh), and therefore is initially one before there is any tendency for arching.

Because the analysis was controlled by reducing σ_s, corresponding settlement at the base of the embankment increased from zero at the edge of the pile cap to a maximum value at the midpoint between pile caps (see Sect. 2.1, Fig. 2.1). The maximum value at the midpoint will now be referred to as δ_s (subsoil), and is thus slightly different to the definition in the ground reaction curve where δ remains consistent with use of fonts is constant for the underground structure (see Sect. 2.3, Fig. 2.13a). This settlement is normalised by the clear spacing between the pile caps ($s - a$), which is equivalent to the width of the structure. Data points are shown at the values of σ_s generated by automatic incrementation in Abaqus, and thus become more dense towards the end of the analysis as plasticity is more prevalent and there is more difficulty in achieving convergence.

The GRC curve is modelled up to the point of maximum arching. Using displacement (rather than stress) controlled analyses it was found that at large displacements a constant value of σ_s was observed, rather than the subsequent increase exhibited in Fig. 2.15 (in Sect. 2.3). It was concluded that the post-maximum stage of the GRC would only be observed in the finite element analyses if brittle soil behaviour was modelled, and it was decided not to introduce such complexity. Nevertheless, the analyses give the stress at maximum arching, and the displacement required to reach this point.

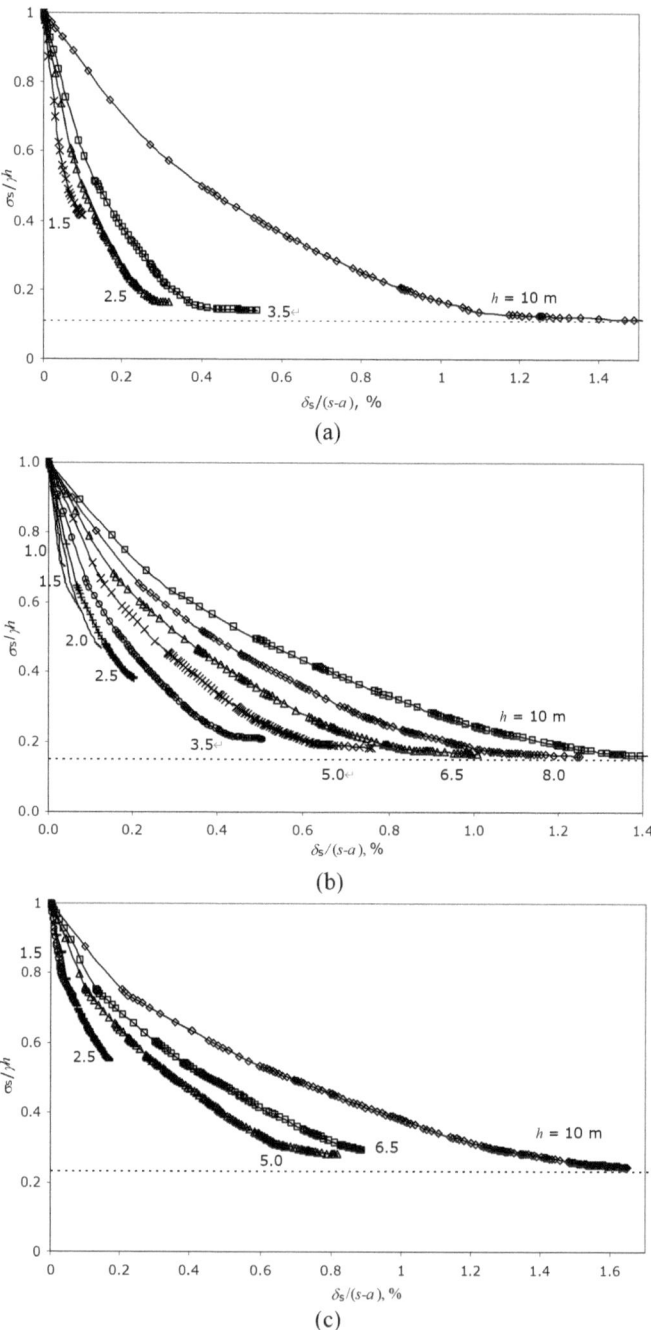

Fig. 3.4 Ground reaction curves for a variety of embankment heights (h); **a** $s = 2.0$ m, **b** s $= 2.5$ m, **c** $s = 3.5$ m, **d** $s = 2.5$ m, $\phi' = 40°$, **e** $s = 2.5$ m, $\psi = 22°$

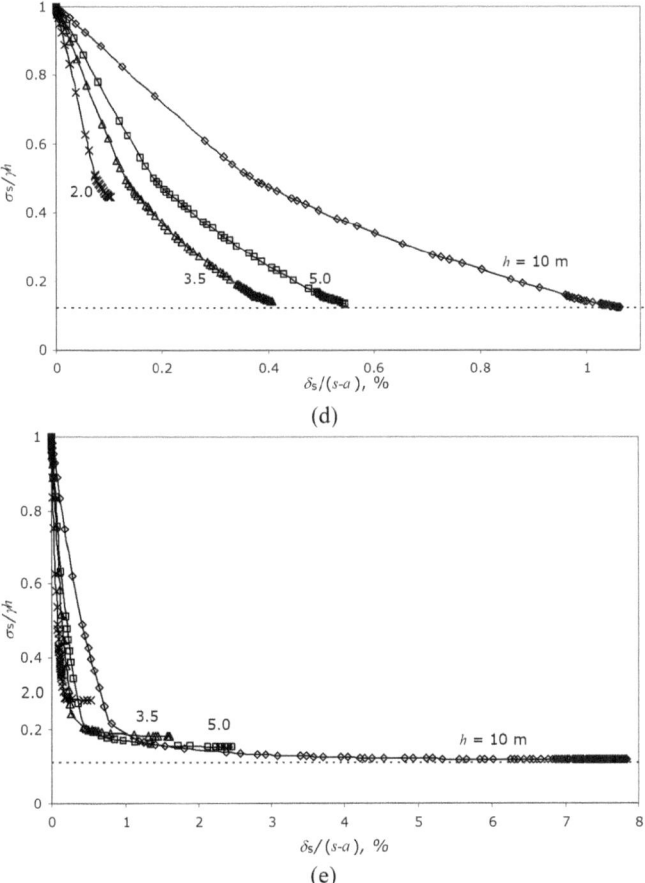

Fig. 3.4 (continued)

Figure 3.4b shows results for the standard soil parameters (Table 3.1), $s = 2.5$ m, and the most comprehensive variety of embankment heights (h). The highest embankment (10 m) requires the largest displacement to reach the point of maximum arching, but even here the normalised displacement is only slightly larger than 1%. However, this value is directly related to the soil stiffness which has been chosen—as anticipated the value was doubled for an analysis with half the soil stiffness. The ultimate normalised stress is in the range from 16 to 20% for $h \geq 3.5$ m, but tends to increase rapidly as h reduces below this value.

Subplots (a) and (c) ($s = 2.0$ and 3.5 m respectively) show trends of behaviour which are similar to (b). The normalised stress at the point of maximum arching for high embankments increases with s.

Subplots (d) and (e) show the effect of increased friction angle and non-zero dilation angle respectively for $s = 2.5$ m. The data again show similar trends. The normalised stresses at the point of maximum arching are slightly lower than that for the standard soil parameters when $s = 2.5$ m. The non-zero dilation angle seems to improve the convergence of the solution towards the end of the analysis, allowing it to continue too much larger displacement at approximately constant subsoil stress. This is probably because the yielding behaviour is closer to an assumption of normality. However, the point of maximum arching is still initially reached at a normalised displacement of less than about 2%. These results will be discussed further in the following Chapter.

3.3.1.2 Midpoint Profile of Earth Pressure Coefficient

It was found that the earth pressure coefficient ($K = \sigma_h'/\sigma_v'$) plotted on a vertical profile at the midpoint between piles (the right-hand boundary of the mesh in Fig. 3.1) gave a good 'illustration' of arching behaviour. Figure 3.5 shows the profiles plotted with z—vertical distance upwards from the base of the embankment, normalised by s. The profiles as plotted do not extend to the top of the embankment for the higher embankments. Values of $0.5(s - a)$, $0.5s$ and $1.5s$ are highlighted on the z axis; and $K = K_0$ and $K = K_p$ (the passive earth pressure coefficient, taking the standard Rankine value and ignoring the small cohesive element of strength) on the K axis. Subplots (a)–(c) again show variation of s whilst (d) and (e) show the effect of increased friction angle and non-zero dilation angle respectively.

Referring to Fig. 3.5b for $(z/s) > 1.5$, $K = K_0$, and thus has not been modified by the formation of the arch. For embankments where $(h/s) > 1.5$, K increases with depth for $z/s < 1.5$, reaching K_p when $z \approx 0.5(s - a)$. Comparing this with a semicircular arch (see Sect. 2.1, Fig. 2.1), the upper limit of the effect of arching is about 3 times higher, but the passive limit is only reached at the inner radius (and below) the arch, where the 'infill' material is evidently in a plastic state.

For embankments where $(h/s) < 1.5$ there is increasing tendency for the highest value of K to occur at the surface of the embankment, initially giving an 'S-shaped' profile, and then monotonic reduction in K with depth in the embankment for the lowest h. In fact, K_p as indicated on the plots neglects the small cohesion intercept, and thus can be exceeded, particularly when stress is small (e.g. near the surface of the embankment or immediately above the subsoil).

Subplots (a) and (c) ($s = 2.0$ and 3.5 m respectively) show trends of behaviour which are similar to (b). When $s = 3.5$ m there is some reduction in K at $z = 0.5(s - a)$ for the largest h, perhaps reflecting an increased tendency for failure of the arch at the pile cap rather than the 'crown' (top of arch) for large s (Hewlett & Randolph, 1988). This trend is supported for $s = 2.0$ m, where there would be increased tendency for passive failure at the crown, and where K is high at and below $z = 0.5(s - a)$.

Subplots (d) and (e) show the effect of increased friction angle and non-zero dilation angle respectively. The data again show similar trends. The higher K_p for the increased friction angle is only fully mobilized when h is close to the 'critical value' of $1.5s$, and K is generally quite considerably less than K_p for $z = 0.5(s - a)$, particularly for the higher embankments. The non-zero dilation angle slightly promotes the tendency for K_p (for the standard friction angle) to be mobilised when $z > 0.5(s - a)$ compared with subplot (b), and the data shows less fluctuation with depth in the plastic infill zone for $z < 0.5(s - a)$. This probably again reflects improved numerical stability in the analysis when plastic strains show a greater degree of normality.

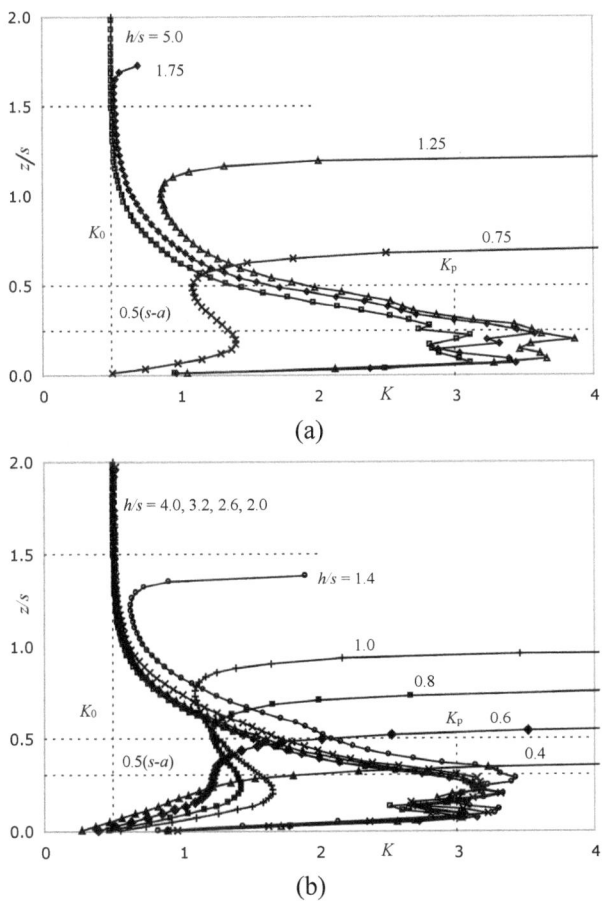

Fig. 3.5 Profiles of earth pressure coefficient (K) on a vertical profile at the midpoint between piles (z measured upwards from base of embankment, see Sect. 2.1, Fig. 2.1), showing variety of embankment heights (h); **a** $s = 2$ m, **b** $s = 2.5$ m, **c** $s = 3.5$ m, **d** $s = 2.5$ m, $\phi' = 40°$, **e** $s = 2.5$ m, $\psi = 22°$

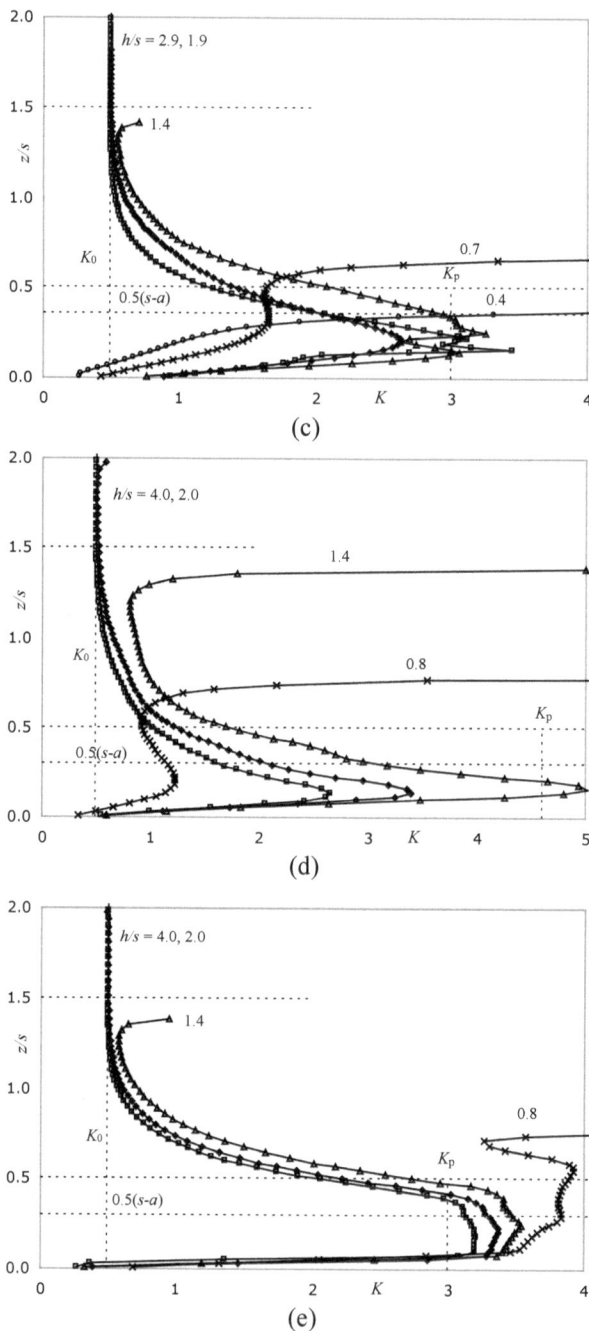

Fig. 3.5 (continued)

3.3.1.3   Ultimate Stress on the Subsoil

Figure 3.6 shows the ultimate stress on the subsoil ($\sigma_{s,ult}$) at the point of maximum arching, illustrating variation with (h/s). Subplot (a) shows normalisation of $\sigma_{s,ult}$ by γh (as in the GRC), whilst (b) shows normalisation by γs.

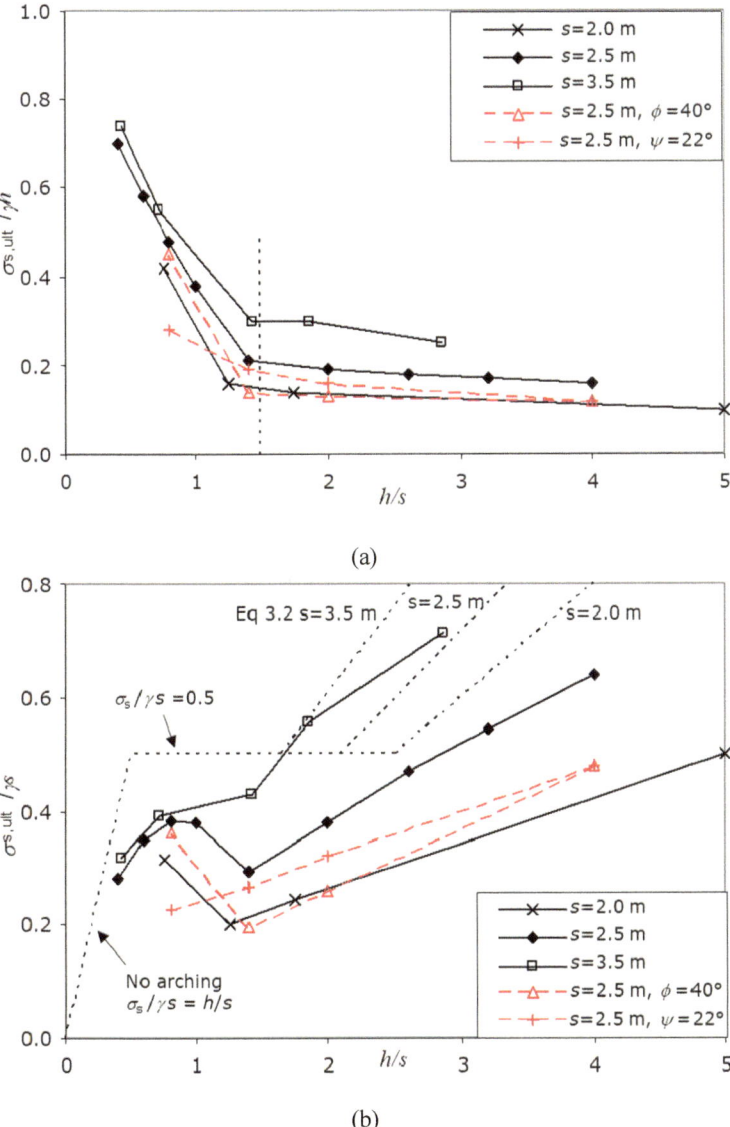

Fig. 3.6 Normalised stress on the subsoil at ultimate conditions ($\sigma_{s,ult}$) showing variation with (h/s); **a** normalised by γh, **b** normalised by γs

Figure 3.6a shows that for $(h/s) > 1.5$, $(\sigma_{s,ult}/\gamma h)$ reduces slowly as h increases, but when $(h/s) < 1.5$, $(\sigma_{s,ult}/\gamma h)$ increases rapidly, tending towards 1.0. This behaviour was previously noted in Fig. 3.4. Also as previously noted the minimum value of $(\sigma_{s,ult}/\gamma h)$ tends to increase with the increase of s, and for $s = 2.5$ m is slightly reduced for increased friction angle or dilation angle.

Figure 3.6b shows lines $\sigma_s = \gamma h$ (i.e. 'no arching'), and $\sigma_{s,ult} = 0.5\gamma s$. Also shown is a simplified version of the condition for failure of the arch at the pile cap proposed by Hewlett and Radolph (1988). The equation of vertical equilibrium for the plane strain situation, assuming σ_s and σ_c (the vertical stress on the subsoil and pile cap respectively) to be constant is:

$$\sigma_c a + \sigma_s(s - a) = \gamma h s \tag{3.1}$$

It is then assumed (from analogy with bearing capacity) that $\sigma_c = K_p{}^2\sigma_s$, to give:

$$\frac{\sigma_s}{\gamma s} = \frac{h}{s} \, \frac{1}{(a/s)\left(K_p^2 - 1\right) + 1} \tag{3.2}$$

This result is plotted for the three values of s.

For small (h/s), $(\sigma_{s,ult}/\gamma s)$ is less than 0.5, and when $(h/s) \approx 0.5$ the data converge with the 'no arching' line. At large h Eq. (3.2) shows the correct trend of behaviour, but tends to overestimate $\sigma_{s,ult}$, particularly as s reduces.

3.3.1.4 Settlement at the Subsoil and Surface of the Embankment

Figure 3.7a shows the maximum value of subsoil settlement at the midpoint between piles (see Sect. 2.1, Fig. 2.1) required to reach ultimate conditions: $\delta_{s,ult}$. This value has been estimated 'by naked eye' from plots such as Fig. 3.4, and thus is somewhat subjective. The value has been normalised by the clear gap between pile caps $(s - a)$ so that it is analogous to $\delta*$ for the GRC (see Sect. 2.3, Fig. 2.15). Variation with (h/s) is shown.

The clearest trend is that the normalised displacement to reach ultimate conditions increases with (h/s), tending to zero when $(h/s) \approx 0.5$, also corresponding to the point of convergence with the 'no arching' line in Fig. 3.6b. If there is no arching then no displacement is required to reach this 'ultimate' condition. This is also evident in Fig. 3.4, where there is less tendency for an ultimate 'plateau' at lower h. As h increases arching occurs, and the amount of stress redistribution from the subsoil to the pile cap increases, thus it is not surprising that the amount of displacement required to achieve ultimate arching conditions also increases. This observation is also consistent with variation with s, which indicates more displacement as s increases (for a given h) since this also implies increased redistribution of load from the subsoil to the pile cap.

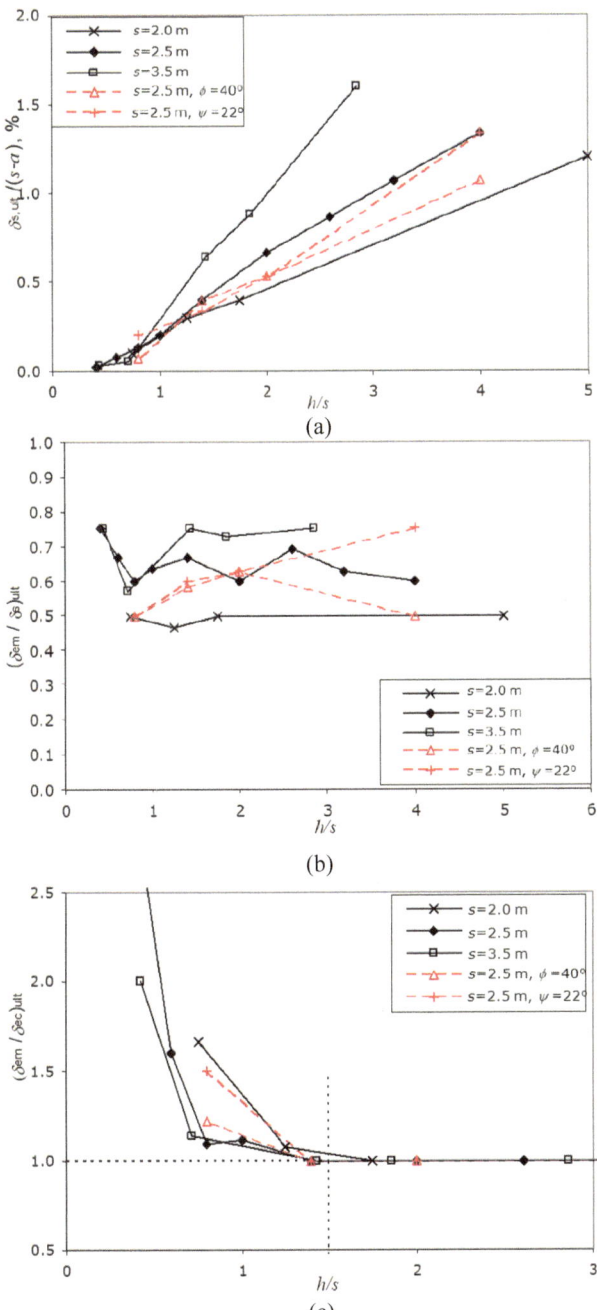

Fig. 3.7 Settlement results at the subsoil and surface of the embankment; **a** ultimate settlement of the subsoil at the midpoint between piles ($\delta_{s,ult}$) normalised by the clear gap between pile caps ($s - a$), **b** ratio of the settlement at the top of the embankment at midpoint between piles (δ_{em}) to the equivalent value in the subsoil (δ_s), **c** ratio of the settlement at the top of the embankment at midpoint between piles (δ_{em}) to the equivalent value at the centreline above the pile cap (δ_{ec})

The absolute magnitude of $\delta_{s,ult}/(s - a)$ is somewhat smaller than the value of $\delta*$ of 2–6% quoted for 'maximum arching' by Lglesia et al. (1999), which appears to be relevant to $(h/s) \approx 2$–5 in the reference. However, the finite element analyses reported here are linear elastic, and the initial gradient of the GRC (Sect. 2.3) implies that the ultimate arching conditions would be reached at a lower value of about 1%. Furthermore, the values shown in Fig. 3.7a would vary directly in inverse proportion to the value of Young's Modulus used in these analyses. For instance, if the Young's modulus had been reduced by a factor of 2 to better simulate the secant modulus to failure the normalised displacement would be doubled.

As shown in Fig. 3.7b the ratio of settlement at the top of the embankment at the midpoint between plies (δ_{em}, see Sect. 2.1, Fig. 2.1) to the equivalent value in the subsoil (δ_s) at the point where ultimate conditions are reached: $(\delta_{em}/\delta_s)_{ult}$, was typically in the range 0.45–0.75. These values seem reasonable: the settlement at the surface of the embankment is less than beneath the arch, but the ratio tends to increase with s.

Figure 3.7c shows the ratio of δ_{em} to the equivalent value at the centreline above the pile cap (δ_{ec}, see Sect. 2.1, Fig. 2.1) at the point where ultimate conditions are reached: $(\delta_{em}/\delta_{ec})_{ult}$, showing variation with (h/s). This is a measure of differential settlement at the surface of the embankment, which is of considerable practical importance in terms of piled embankments. For $(h/s) > 1.5$ the value is 1.0, indicating no differential settlement. As h reduces below this value, differential settlement increases, dramatically so for (h/s) is less than about 0.75.

3.3.2 For Three-Dimensional Condition

3.3.2.1 Ground Reaction Curves

Figure 3.8 shows the three-dimensional results plotted as a 'ground reaction curve' (GRC). Normally the stress acting on the 'roof' of the structure would be considered, but here the subsoil stress (σ_s) is used. It is normalised by the nominal overburden stress due to the embankment γh, giving a value of 1.0 at the start of the analysis before there is any arching. The data points show the automatic incrementation in Abaqus, as σ_s was reduced to stimulate arching. The x-axis shows the maximum settlement at the base of the embankment (i.e. the subsoil settlement at point A), δ_s, normalised by the clear gap between adjacent pile caps ($s - a$).

Results for a range of embankment heights (h) from 1.0 to 10.0 m for analyses with pile spacing $s = 2.5$ m are shown in Fig. 3.8b. The initial shape of the GRCs are as reported by Iglesia et al. (1999) with σ_s reducing quite rapidly at small δ_s. Apart from very low embankment heights, an ultimate point of 'maximum arching' is reached at normalised displacement between 1 and 4%, tending to increase with h. The Young's modulus chosen for the embankment soil (Table 3.1) is plausible for a simple linear elastic approach, but is somewhat arbitrary, and as expected the amount

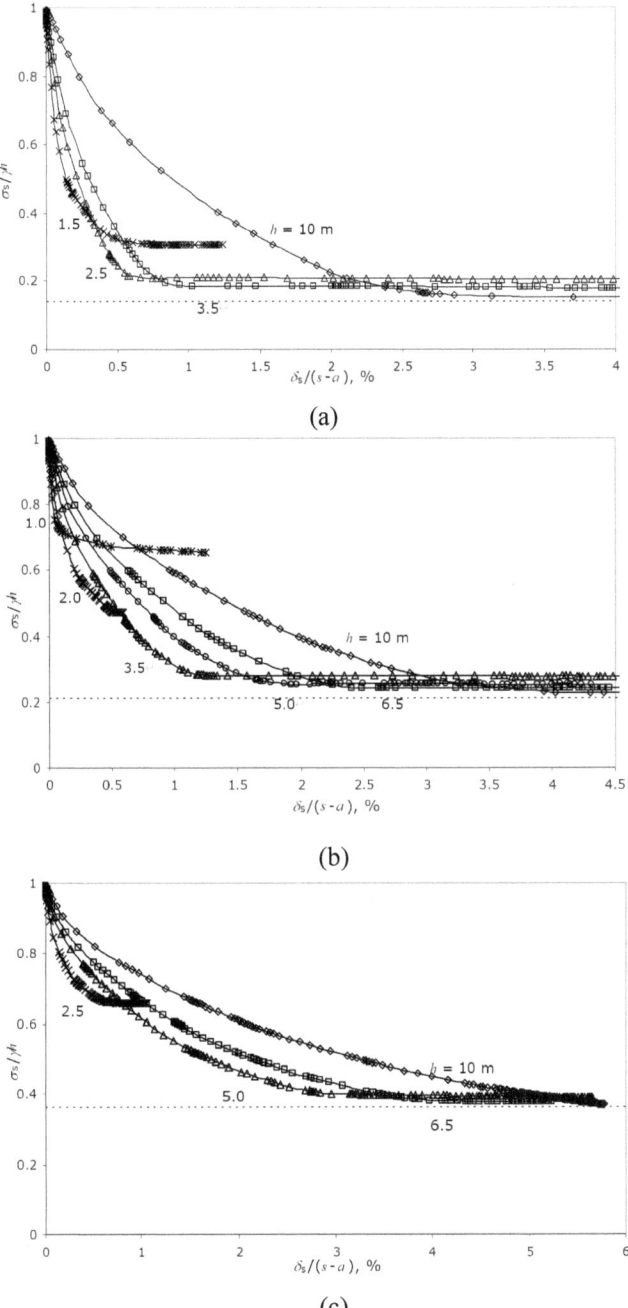

Fig. 3.8 Ground reaction curves for a variety of embankment heights (*h*); **a** *s* = 2.0 m, **b** *s* = 2.5 m, **c** *s* = 3.5 m

of strain required to reach the point of maximum arching was inversely proportional to the soil stiffness.

Subplots (a) and (c) ($s = 2.0$ m and 3.5 m respectively) show trends of behaviour which are similar to (b). The normalised stress at the point of maximum arching increases with the rise of s, which is consistent with behaviour in the plane strain situation.

The normalised displacement at the point of maximum arching is about three times larger than that obtained in the equivalent plane strain analyses. As indicated, the points of maximum arching tend towards a unique value of about 0.2 for high embankments.

3.3.2.2 Midpoint Profile of Earth Pressure Coefficient

Figure 3.9 shows the earth pressure coefficient (K) at the point of maximum arching on a vertical profile through points A and B respectively (Fig. 3.2), for a range of values h and s. z is the height above the base of the embankment. At both points the horizontal stress was found to be independent of direction, and K is derived using the actual (not nominal) vertical stress.

The Rankine active and passive values of K are indicated on the plots. As discussed by Zhuang et al. (2010), owing to the small cohesion intercept used for the material strength (Table 3.2) it is possible for these 'limits' to be exceeded, particularly near the surface of the embankment. Values of $z = s/\sqrt{2}$ and $(s - a)/\sqrt{2}$ are also indicated in Figs. 3.9 and 3.10, corresponding to the outer and inner radii of a three-dimensional hemispherical arch as proposed by Hewlett and Randolph (1988), at the highest point ('crown').

The three-dimensional results in Fig. 3.9 can be compared with corresponding plane strain analyses, and show similar general patterns of behaviour. $K = K_0$ (i.e. no effect from arching) is observed at a height about twice that implied by the top of the hemispherical arch. However, K increases quite slowly with depth, only reaching K_p at the inner radius of the hemisphere. Lower embankments show first an 'S'-shaped profile and then monotonic increase in K with height in the embankment.

Figure 3.9d shows an equivalent plot for point B (at the centre of a pile cap). When $z/(s - a) < 1.5$ an active state is observed for the higher embankments, and K returns to K_0 at a similar height to Fig. 3.9b.

The results observed in Fig. 3.9 are consistent with the three-dimensional hemispherical arch as proposed by Hewlett and Randolph (1988) in the sense that active conditions are observed above the pile cap, and passive conditions are observed at the 'crown' of the arch (at least near the inner radius). These conditions are associated with 'punching' of the pile caps into the base of the embankment and failure of the arch at the mid-span respectively. It is observed that the effect on the stress state throughout the soil is more widespread (higher) than the proposed discrete hemispherical boundaries, since there is a gradual (rather than instantaneous) transition back to K_0 as height increases.

Fig. 3.9 Profiles of earth
pressure coefficient at the
point of maximum arching
with height above base of the
embankment (z) for various
embankment heights (h) and
centre-centre pile spacing s

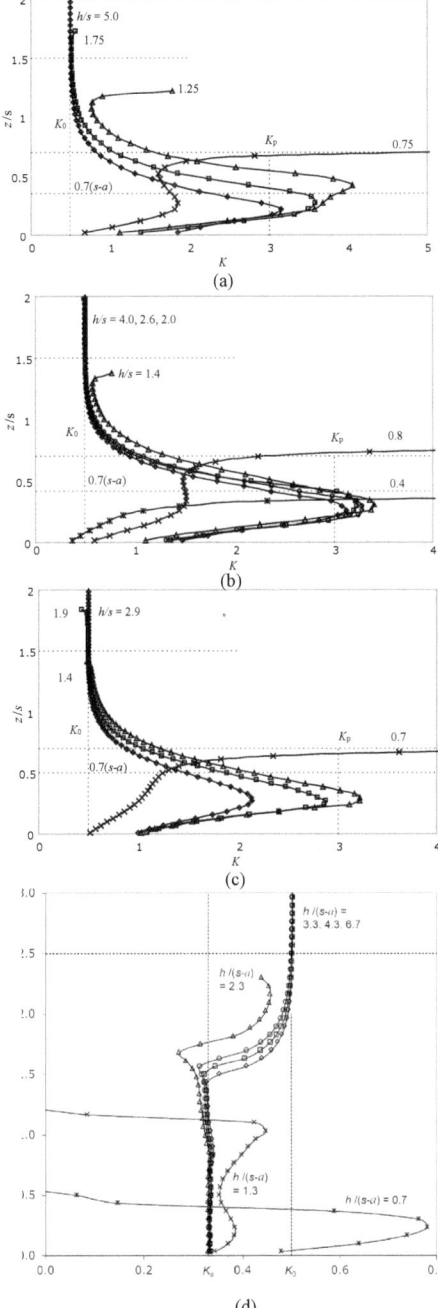

Fig. 3.10 Geometry of arching in the three-dimensional condition

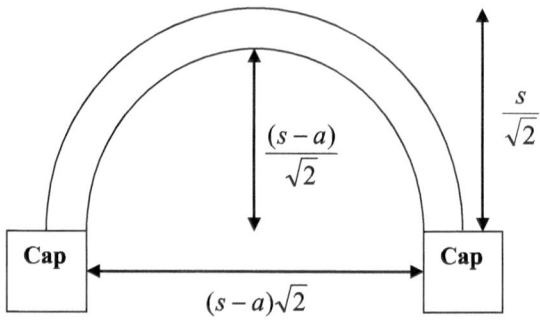

$$\frac{(s-a)}{\sqrt{2}}$$

$$\frac{s}{\sqrt{2}}$$

Cap

Cap

$$(s-a)\sqrt{2}$$

Table 3.4 Height of arching above a void or in a piled embankment reported by various authors

	Terzaghi (1943)	Potts and Zdravkovic (2008)	Ellis and Aslam (2009b)	Britton and Naughton (2010)	Zhuang et al. (2010) and in this chapter
Normalising length	Void width	Void width	$(s-a)$	$(s-a)$	$(s-a)$
Plane strain	2–3	3	NA	NA	3[*]
3-d	NA	NA	2	1.94–3.1	2.5

[*] Based on alternative normalising variable compared to the original paper

Table 3.4 summarises similar results for the height of influence of arching over a void and piled embankment in plane strain or three dimensions. The data are based on results published by Terzaghi (1943), Potts and Zdravkovic (2008), Ellis and Aslam (2009b) and Britton and Naughton (2010). In the case of a piled embankment $(s - a)$ is directly equivalent to the void width for plane strain. There is reasonable consistency throughout the results. Data are relatively limited for the three-dimensional case, but indicate that the normalised height of influence is actually slightly lower compared to plane strain.

3.3.2.3   Ultimate Stress on the Subsoil

Figure 3.11 shows the subsoil stress (σ_s) at the point of maximum arching in each analysis normalised by $\gamma(s - a)$, showing variation with normalised embankment height $h/(s - a)$, as proposed by Ellis and Aslam (2009a). Results for each value of s are shown as separate data series joined by solid lines. The three-dimensional results in Fig. 3.11 can be compared with corresponding plane strain analyses (Zhuang et al., 2010), and show similar trends. A number of comparison lines are shown: 'no arching' ($\sigma_s = \gamma h$), a flat section and predictions of failure at the pile cap as proposed by Hewlett and Randolph's three-dimensional method (for $\phi = 30°$).

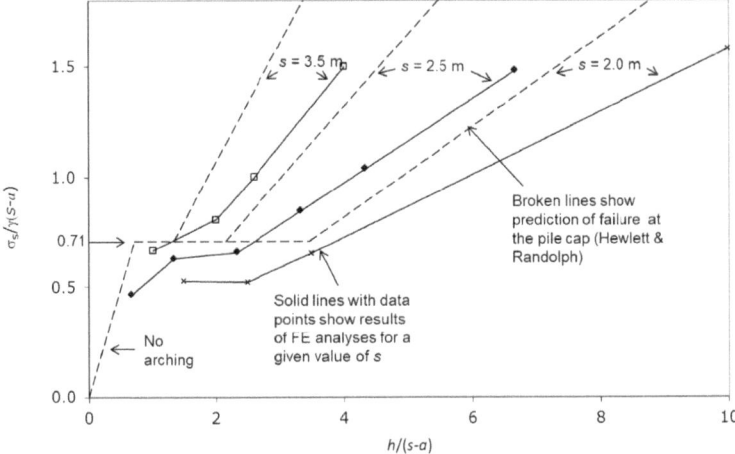

Fig. 3.11 Variation of behaviour at point of maximum arching showing variation with normalised embankment height

The flat section indicates $\sigma_s = \gamma(s - a)/\sqrt{2} = 0.71\gamma(s - a)$. This corresponds to the nominal weight beneath the crown of the hemisphere, and appears in the corresponding formula for failure at the crown proposed by Hewlett and Randolph (1988). In Fig. 3.11, $\sigma_s/\gamma(s - a) = 0.71$ is a reasonable upper bound to the data for low embankments. It can be compared with a value of 0.50 proposed by Ellis and Aslam (2009a), based on the results of three-dimensional centrifuge tests. The higher value for the numerical modelling reported here is potentially attributable to the friction angle of 30° assumed, which would be quite conservative compared to the dense sand used in the centrifuge tests.

Proceeding to higher embankments and the inclined lines for predicted failure at the pile cap, the FE analyses show a consistent trend of behaviour, but with lower σ_s. However, the relevance of this failure mode is confirmed.

3.3.2.4 Settlement at the Subsoil and Surface of the Embankment

Figure 3.12a shows the maximum value of subsoil settlement at the centre point of the basic unit (D, Fig. 3.2), normalised by the clear gap between pile caps ($s - a$) showing variation with h/s. As in plane strain there is a clear trend for the settlement at ultimate conditions to increase with (h/s), tending to zero when (h/s) ≈ 0.5 (corresponding to no arching).

As h increases arching occurs, and the amount of stress redistribution from the subsoil to the pile cap increases, thus it is not surprising that the amount of displacement required to achieve ultimate arching conditions also increases. This observation is also consistent with the variation with s, which indicates more displacement as s

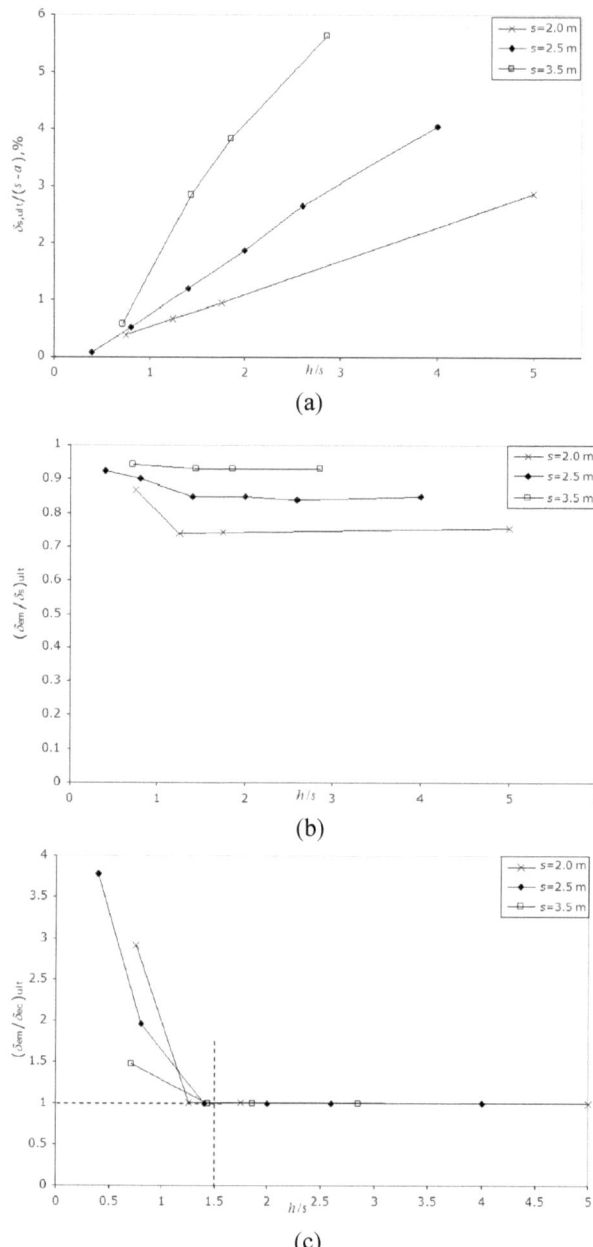

(a)

(b)

(c)

Fig. 3.12 Settlement results at the subsoil and surface of the embankment; **a** ultimate settlement of the subsoil at the midpoint of diagonal between piles ($\delta_{s,ult}$) normalised by the clear gap between pile caps ($s - a$), **b** ratio of the settlement at the top of the embankment at the midpoint of diagonal between piles (δ_{em}) to the equivalent value in the subsoil (δ_s), **c** Ratio of the settlement at the top of the embankment at the midpoint of diagonal between piles (δ_{em}) to the equivalent value at the centre above the pile cap (δ_{ec})

increases (for a given h) since this also implies increased redistribution of load from the subsoil to the pile cap.

The absolute magnitude of $\delta_{s,ult}/(s - a)$ is approximately between 1 and 6% when $(h/s) \approx$ 2–5, this value is considerably larger than equivalent value in plane strain situation (by a factor of about 3). This is because there is less support from the pile caps in the 3D situation, which causes larger displacement at the 'ultimate' condition.

Subplot (b) shows the ratio of settlement at the top of the embankment at the midpoint of the diagonal between piles above point D (δ_{em}, see plane strain graph Fig. 2.1, in Sect. 2.1) to the equivalent value in the subsoil (δ_s) at the point where ultimate conditions are reached: $(\delta_{em}/\delta_s)_{ult}$, showing variation with (h/s). Values range from 0.75 to 0.95. This is larger than the equivalent value in the plane strain situation (0.45–0.75). Also the value is largest, tending to 1.0 for increasing s. For low (h/s) the value of 1.0 most likely indicates no arching. At higher (h/s) the relatively high value of this ratio probably again reflects some reduction in the 'effectiveness' of arching in 3D compared to plane strain condition.

Subplot (c) shows the ratio δ_{em} to the equivalent value at the centre above the pile cap (δ_{ec}, see plane strain graph Fig. 2.1, in Sect. 2.1) at the point where ultimate conditions are reached: $(\delta_{em}/\delta_{ec})_{ult}$, showing variation with (h/s). This graph shows a similar trend to the plane strain situation. For $(h/s) > 1.5$ the value is 1.0, indicating no differential settlement. As (h/s) reduces differential settlement increases, dramatically so for (h/s) less than about 1.4. The three-dimensional results show similar trends to plane strain analyses. As h/s increases, the ratio $(\delta_{em}/\delta_{ec})_{ult}$ drops to 1.0, implying no differential settlement when h/s is in the range 1.25–1.50. The upper limit of this range corresponds to $z/(s - a) = 1.5$.

3.4 Summary

The plane strain and three-dimensional FE analyses of arching in a piled embankment have been presented. The concept of the 'ground reaction curve', initially proposed by Iglesia et al. (1999) and applied to plane strain arching, which has also been extended to the three-dimensional situation.

The results for stress on the subsoil at the point of 'maximum arching' are presented in a format proposed by Ellis and Aslam (2009a), and the results are observed to be consistent with the results of centrifuge model tests where data are available. In terms of comparison with existing analytical approaches, the importance of 'punching' of pile caps into the base of high embankments is confirmed. This effect is only explicitly considered in the analytical approach proposed by Hewlett and Randolph (1988). The relevance of the mechanism is also confirmed by the existence of active earth pressure above the pile caps. The results of a series of linearly elastic-perfectly plastic plane strain and three-dimensional finite element analyses to investigate the arching of a granular embankment supported by pile caps over a soft subsoil have been presented and discussed. The analyses demonstrate that the ratio of the embankment height to the centre-to-centre pile spacing (h/s) is a key parameter to determine the performance of the embankment system:

- (h/s) ≤ 0.5 there is virtually no effect of arching: 'ultimate' conditions are reached almost immediately (with very small displacement) in the analysis, relative differential settlement at the surface of the embankment is very large, and the stress acting on the subsoil is virtually unmodified from the nominal overburden stress.
- 0.5 ≤ (h/s) ≤ 1.5 there is increasing evidence of arching: as (h/s) increases the displacement required to reach 'ultimate' conditions increases. Relative differential displacement at the surface of the embankment reduces, and the stress acting on the subsoil reduces compared with the nominal overburden stress.
- 1.5 ≤ (h/s) 'full' arching is observed: the displacement required to reach ultimate conditions continues to increase and a clearly defined ultimate state is maintained at large displacement. There is no differential displacement at the surface of the embankment, and the stress acting on the subsoil is considerably reduced compared with the nominal overburden stress. For a high embankment the stress state is not significantly affected above a height of 1.5 s in the embankment.

The vertical extent of arching in the embankment was found to be 1.25–1.50 times the centre-to-centre pile spacing between pile caps, which is actually slightly lower compared to plane strain. This is consistent with the results from other researchers, and dictates the minimum height of a piled embankment for effective arching to be developed. Furthermore, it has been shown that up to a critical value of (h/s) the stress on the subsoil is less than $0.5\gamma s$, approximately representing the effect of the infill material below the arch. At higher values of (h/s) conditions at the pile cap are critical and Eq. (3.2) can be used to conservatively estimate the stress on the subsoil. As for the differential settlement, the three-dimensional results show similar trends to plane strain analyses. As h/s increases, the ratio (δ_{em}/δ_{ec})$_{ult}$ drops to 1.0, implying no differential settlement occurs when h/s is in the range 1.25–1.50. The upper limit of this range corresponds to $z/(s - a) = 1.5$.

References

Britton, E., & Naughton, P. J. (2010). An experimental study to determine the location of the critical height in piled embankments. In *Proceedings of the 9th international conference on geosynthetics*, Brazil, pp. 1961–1964.

Ellis, E. A., & Aslam, R. (2009a). Arching in piled embankments: Comparison of centrifuge tests and predictive methods—Part 1 of 2. In *Ground engineering* (pp. 34–38), June 2009.

Ellis, E. A., & Aslam, R. (2009b). Arching in piled embankments: Comparison of centrifuge tests and predictive methods—Part 2 of 2. In *Ground engineering* (pp. 28–31), July 2009.

Hewlett, W. J., & Randolph, M. F. (1988). Analysis of piled embankments. In *Ground engineering* (pp. 12–18), April 1988.

Iglesia, G. R., Einstein, H. H., & Whitman, R. V. (1999). Determination of vertical loading on underground structures based on an arching evolution concept. In *Proceedings of 3rd national conference on geo-engineering for underground facilities*.

Naughton, P. J., & Kempton, G. T. (2005). *Comparison of analytical and numerical analysis design methods for piled embankments. Contemporary issues in foundation engineering*, GSP 131, ASCE, Geo-Frontiers, Austin, Texas.

Naughton, P., Scotto, M., & Kempton, G. (2008). Piled embankments: Past experience and future perspectives. In *Proceedings of the 4th European geosynthetics conference*, Edinburgh, UK, September 2008, Paper number 184.

Potts, V. J., & Zdravkovic, L. (2008). Finite element analysis of arching behaviour in soils. In *12th international conference of international association for computer methods and advances in geomechanics (IACMAG)*, Goa, India, October 2008, pp. 3642–3649.

Stewart, M. E., & Filz, G. M. (2005). Influence of clay compressibility on geosynthetic loads in bridging layers for column—Supported embankments. In *Proceedings of GeoFrontiers 2005*, ASCE, Austin.

Terzaghi, K. (1943). *Theoretical soil mechanics*. Wiley.

Zhuang, Y., Ellis, E. A., & Yu, H. S. (2010). Plane strain FE analysis of arching in a piled embankment. *ICE Proceedings (Ground Improvement), 163*(GI4), 207–215.

Chapter 4
Geogrid Reinforced Piled Embankment in Plane Strain and Three-Dimensional Conditions

4.1 Introduction

This chapter considers the performance of piled embankments where one or more layers of geotextile or geogrid reinforcement are used at the base of the embankment, for a wide range of piled embankment geometries. Again the plane strain and three-dimensional models of a geogrid (or geotextile) reinforced piled embankment are respectively adopted, forming a logical extension of the analyses presented in the previous chapter.

A piled embankment (Fig. 4.1a) relies on arching of the embankment material to transfer a significant amount of the embankment load directly onto pile caps. The remaining load (e.g. below the arch) is carried by some combination of tensile reinforcement (which is frequently used) and/or the subsoil (if this is realistic/admissible in design).

The British 'Code of practice for strengthened/reinforced soils and other fills' (BS 8006) was substantially revised in 2010 (BS 8006, 2010). Historically BS 8006 considered arching in a piled embankment based on an interpretation of the 'Marston' equation, and corresponding tension in a single layer of extensible reinforcement near the base of the embankment. The 2010 revision included an alternative method for arching proposed by Hewlett and Randolph (1988), and a new 'minimum limit' on reinforcement tension.

Van Eekelen et al. (2011) proposed modification of BS 8006 for piled embankments based on comparison with axisymmetric Finite Element (FE) analyses. However, the comparison did not consider the Hewlett and Randolph approach, and the use of axisymmetric Finite Element (FE) analysis is questionable. There are other examples in the literature modelling biaxial geogrid reinforcement as an isotropic material (e.g. Liu et al., 2007).

Plaut and Filz (2010), Jones et al. (2010), and Halvordson et al. (2010) presented an interesting comparison of axisymmetric and isotropic geogrid reinforcement respectively. The results indicate considerable variation in the maximum tension, which

© The Author(s) 2025
Y. Zhuang, *Arching in Piled Embankments Including the Effects of Reinforcement and Subsoil*, https://doi.org/10.1007/978-981-96-1198-0_4

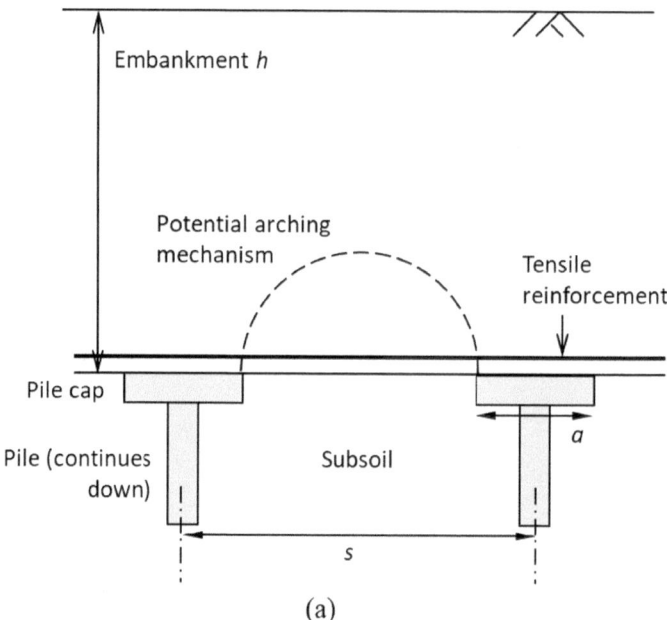

(a)

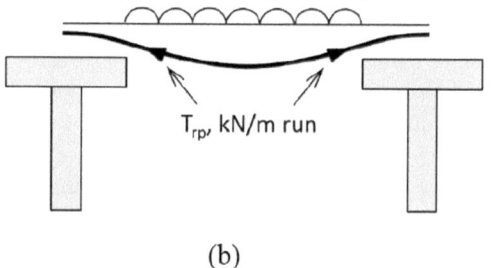

(b)

Fig. 4.1 Section through piled embankment illustrating behaviour and notation; **a** schematic of behaviour and geometry, **b** reinforcement load (W_T) and tension (T_{rp})

is observed at the corner of a pile cap. The isotropic model gives a considerable 'spike' in tension at this point. Conversely the axisymmetric model gives very little increase at the edge of the pile cap (noting that a corner is not actually modelled as such). Whilst these studies give a good insight into different approaches to modelling of the reinforcement, other aspects of the modelling are somewhat idealised. Most notably embankment loading is modelled using an assumed uniform stress, and subsoil support is present (modelled using springs), and carries a significant amount of the embankment load.

The implementation of the Hewlett and Randolph approach in BS 8006 was modified in a 2012 'Corrigendum' (BS8006, 2012). This chapter considers BS 8006 predictions of reinforcement tension derived from the 'Marston' and 'Hewlett and Randolph' approaches as a basis for embankment load on the reinforcement, for a wide range of piled embankment geometries. The predictions are compared with the results obtained from three-dimensional FE analyses. In keeping with BS 8006, concepts of reinforcement behaviour are not further extended to separate orthogonal layers of reinforcement (e.g. Love & Milligan, 2003). Predictions of maximum sag for the reinforcement are also considered. The effect of the side slopes of the embankment are not considered (noting that this is considered separately in BS 8006). Corresponding to the approach adopted in BS 8006, it is assumed that there is no support from the underlying subsoil.

4.2 Calculation of Reinforcement Tension in BS8006 (2010)

In this section, the method to estimate the reinforcement tension of geogrid T_{rp} based on the BS8006 approach was exhibited and discussed. To calculate the reinforcement tension, the methods to estimate the nominal vertical stress acting on the reinforcement σ_r considering soil arching effect, the subsoil support, and the distributed load acting on the extensible reinforcement W_T were presented, respectively.

4.2.1 Embankment Arching

Key geometrical variables in consideration of arching are (Fig. 4.1a): a, the pile cap size; s, the centre-to-centre pile spacing; h, the embankment height. Here the pile geometry will refer to square caps on a square grid (i.e. a three-dimensional situation), referred to as the 'most commonly used in practice' in BS 8006. Historically BS 8006 considered arching in a piled embankment based on an interpretation of the 'Marston' equation for 'projecting subsurface conduits'. The 2010 revision included an alternative method for arching proposed by Hewlett and Randolph (1988), based on specific consideration of arching in a piled embankment.

Hewlett and Randolph introduced the concept of or arching 'efficacy', E, referred to as 'efficiency' in BS 8006. E is defined as 'the proportion of the embankment weight carried by the piles, hence the proportion carried by the geosynthetic reinforcement may be determined as $(1-E)$' (BS 8006). In fact, this statement will be qualified here as 'the weight transferred to the piles directly by arching', since the remaining weight carried by the reinforcement is also ultimately transferred to the piles. Furthermore, there are two methods for calculation of E (failure at the 'crown' of the arch or pile cap), and hence the minimum value E_{min} is used.

Here, a nominal vertical stress acting on the reinforcement, σ_r, will be defined (noting that this is not explicitly used in BS 8006). It follows from vertical equilibrium (with no subsoil support) that

$$\sigma_r = (\gamma h + w_s)(1 - E_{min})\frac{s^2}{s^2 - a^2} \qquad (4.1)$$

Omitting partial load factors f, γh is the surcharge at the surface (taken as zero in this study), both acting on plan area s^2. The reinforcement vertical stress acts on area $(s^2 - a^2)$ Love and Milligan (2003) give formulae for the reinforcement vertical stress σ_r which avoid ('bypass') the use of E.

4.2.2   Subsoil Support

Although complete loss of support from the subsoil may be considered conservative under some circumstances, while it is robust, and implicit in BS 8006. However, this assumption can cause difficulty in comparison with field studies. For instance, Briancon and Simon (2012) report a field study where the compressibility of the subsoil 'was not very high', and consequently the results indicate that the subsoil carries quite significant load, whilst the reinforcement strain is less than 1%. Almeida et al. (2007) described a field study where excavations were performed under the geogrid to remove subsoil support. However, the embankment height was low (1.3 m).

As noted by Van Eekelen et al. (2011) 'a considerable degree of subsoil support has been measured in the available field studies … Although this support was measured during monitoring, it may disappear in future years'. The long-term loss of support will most obviously be associated with dissipation of initial excess pore pressures (i.e. consolidation and associated settlement). There are also potential issues with future fluctuations in the groundwater table, or the presence of imported fill material below pile cap level, which will load the subsoil directly (unaffected by arching above the pile caps).

4.2.3   Determination of W_T

In the absence of subsoil support the tensile reinforcement will be required to carry any load not transferred to the pile caps by arching in the embankment (e.g. below the arch in Fig. 4.1a). Thus, determination of the reinforcement load in BS 8006 begins with consideration of embankment arching.

BS8006 (2012) presents three methods for the calculation of the distributed (vertical) load acting on the extensible reinforcement (W_T, kN/m; Fig. 4.1b), by way of:

(a)   Marston's formula
(b)   Hewlett and Randolph's (1988) approach
(c)   A minimum value (W_{Tmin}).

Both (a) and (b) were referred to in the section on embankment arching above. In fact, the predictions do not relate exclusively to the named methods, since they are only used to consider arching in the embankment, and additional assumptions (below) are required to proceed to an assessment of reinforcement tension. Nevertheless, it will be convenient to refer to the methods using this nomenclature. Method (c) was introduced in BS8006 (2010).

The detailed implementation of the formulation for W_T in BS 8006 has been the subject of discussion beyond assumptions related to arching behaviour in the embankment and loss of subsoil support. For instance, W_T will not be uniform. Also, how is the vertical load from the arching embankment 'distributed' as load on the reinforcement (Love & Milligan, 2003)? Van Eekelen et al. (2011) argued that the approach used in BS8006 can be modified to reduce W_T in this respect, and this appears to reflect the BS8006 2012 Corrigendum as applied to the Hewlett and Randolph (1988) approach.

Using the nominal vertical stress acting on the reinforcement (σ_r) defined in Eq. 4.1, the equations for W_T based on the Hewlett and Randolph approach in the 2010 and 2012 versions of BS8006 can be written as:

$$W_T = s\sigma_r \tag{4.2a}$$

$$W_T = \tfrac{1}{2}(s + a)\sigma_r \tag{4.2b}$$

respectively.

Given that' $s > a$, the more recent approach is less conservative (e.g. a 30% reduction in is implied for $s/a = 2.5$). Use of the Hewlett and Randolph (1988) method to determine σ_r does not change, but W_T in BS 8006 is affected by the 2012 Corrigendum. Equations 4.2a, b correspond to those proposed by Love and Milligan (2003) for 'transverse (lower layer)' and 'average' allocations of vertical loading on orthogonal layers of reinforcement, respectively. Love and Milligan also proposed that W_T for 'longitudinal' (upper layer) reinforcement could be calculated as a σ_r, noting that this would be less than the lower layer. Thus vertical equilibrium is satisfied by W_T corresponding to Eq. 4.2b in both directions, but not satisfying Eq. 4.2a. Equation 4.2a is not current in BS 8006 and does not correspond correctly with vertical equilibrium as an average reinforcement value. Nevertheless, it will be considered here since the updated version in Eq. 4.2b is less conservative.

4.2.4   Determination of T_{rp}

Once W_T has been determined, BS 8006 () provides a formula for calculation of tension in an extensible reinforcement T_{rp}:

$$T_{rp} = W_T \tfrac{s-a}{2a}\left(1 + \tfrac{1}{6\varepsilon}\right)^{0.5} \tag{4.3}$$

There are two unknowns: T_{rp} is the tensile load in the reinforcement (kN/m) and ε is the strain mobilized in the reinforcement.

As strain (and sag) increases the ratio T_{rp}/W_T reduces. BS8006 states that the equation may be solved by 'the maximum allowable strain in the reinforcement and by an understanding of the load/strain characteristics of the reinforcement.' Normally it will be possible to assign the reinforcement an appropriate long-term secant stiffness (J, MN/m), which should make due allowance for creep. Then T_{rp} = J, allowing solution of the equation:

$$T_{rp} = W_T \tfrac{s-a}{2a}\left(1 + \tfrac{J}{6T_{rp}}\right)^{0.5} \tag{4.4}$$

Given that T_{rp} appears on both sides of the equation some iteration is required, but this can be easily automated on a personal computer. In fact, for typical strains $(1/6\varepsilon)$ is quite large compared to 1 (but less so as ε increases), and the equation can be approximated as

$$T_{rp} \approx 0.35J^{1/3}\left[W_T\left(\tfrac{s-a}{a}\right)\right]^{2/3} \tag{4.5}$$

This approximation is not conservative in estimation of T_{rp}). However, it may be useful for a rough estimate not requiring iteration. A corresponding estimate of strain is

$$\varepsilon = \tfrac{T_{rp}}{J} \approx 0.35\left[\tfrac{W_T}{J}\left(\tfrac{s-a}{a}\right)\right]^{2/3} \tag{4.6}$$

This reveals that (based on this approach) tension and strain vary as $W_T{}^{2/3}$, whilst strain in the reinforcement varies as $(1/J)^{2/3}$ (noting that W_T is independent of J).

4.3 Finite-Element Analyses

The numerical models of piled embankment reinforced with one or more layers of geotextile or geogrid at the base of the embankment for the plane strain conditions and three-dimensional condition were introduced in this section. The material parameters for geogrid, finite element mesh, boundary conditions, and summary cases all reported in detail.

4.3.1 For Plane Strain Condition

Figure 4.2 shows a typical mesh for the reinforced embankment, with height $h = 3.5$ m and pile spacing $s = 2.5$ m. There are 1566 eight nodded, reduced-integration, two-dimensional, quadratic solid elements (CPE8R) for the embankment. The geogrid is

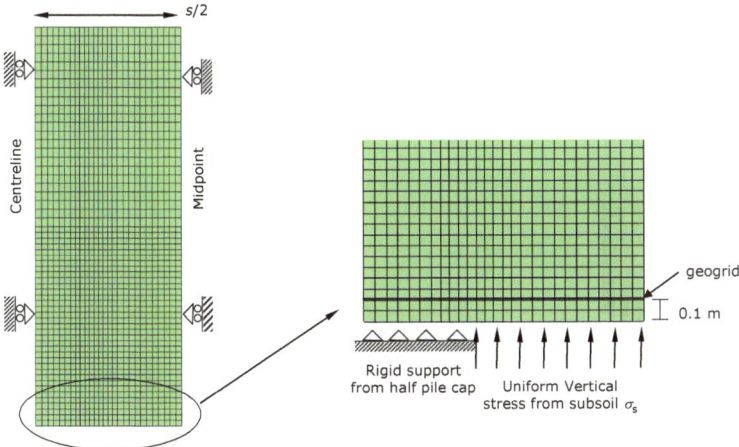

Fig. 4.2 Typical finite element mesh ($h = 3.5$ m, $s = 2.5$ m, one layer of reinforcement) and boundary conditions for reinforced embankment

modelled using 16 three noded quadratic truss elements (T2D3). Truss elements are used to model slender, line-like structures that carry loading only along the axis or the centreline of the element, and no moments or forces perpendicular to the centreline. Hence the tensile action of the reinforcement is modelled with no bending stiffness.

A contact model is required to 'join' the 'solid' elements used for the soil and the 'truss' elements used for the geogrid. A 'surface to surface' contact type has been used. As in the previous Chapter, the vertical boundaries represent lines of symmetry at the centreline of a support (pile cap), and the midpoint between supports (see Sect. 2.1, Fig. 2.1) with corresponding restraint on horizontal movement. The reinforcement is positioned 0.1 m above the base of the embankment, and there is likewise restraint on horizontal movement at both ends. In some analyses three layers of reinforcement were used at 0.1, 0.4 and 0.7 m above the base of the embankment, in an attempt to simulate a 'Load Transfer Platform' (LTP).

Again there are no boundary conditions imposed at the top embankment surface, and no surcharge is considered to act here. The bottom boundary represents the base of the embankment, with the same boundary conditions as used in the previous chapter. The vertical stress in the subsoil supporting the embankment (σ_s) is again used to control the analysis—the subsoil itself was not actually modelled. As the subsoil stress (σ_s) reduces, arching will occur as studied in the previous Chapter, but tension will also be generated in the geogrid.

The pile cap width (a) was again fixed at 1 m and the centre-to-centre spacing (s) was 2.0, 2.5 or 3.5 m. The embankment height (h) was 1, 3.5 or 10 m. Throughout the analyses the minimum and maximum element sizes of the embankment were approximately 0.0006 and 0.00738 m^3/m. This corresponds to side lengths of approximately 30–150 mm. The truss element lengths were 20–100 mm.

The embankment material was again assumed to be granular, and modelled using the standard linear elastic and Mohr Coulomb yield criterion parameters as the previous Chapter. The parameters used for the reinforcement are shown in Table 4.1. The tensile stiffness J was 6 or 12 MN/m for a single layer of reinforcement, or 2 MN/m for each of the layers where 3 layers were modelled. This is effectively specified using the product of the Young's Modulus of the material and the cross-sectional area per metre width. Here the latter was taken as a nominal 1 mm thickness, and corresponding values of Young's Modulus were assigned to give the required value of J. The Poisson's Ratio of zero means that axial strain does not affect the plane strain direction. The interface friction angle (δ_i) between the embankment fill and geogrid was 0° or 20°, corresponding to a nominally 'smooth' or 'rough' interface.

All analyses reported in this Chapter are summarised in Table 4.2. Variations to the 'standard' parameters are highlighted in bold. In the first three analyses, for $s = 2.5$ m and $J = 6$ MN/m with $\delta_i = 0$, the effect of increasing h from 1 to 10 m was considered. Then for $h = 3.5$ m, the influence of increasing s from 2.0 to 3.5 m was considered.

The remaining analyses considered the effect J, δ_i and the number of geogrid layers (N). Where three layers of geogrid were used, it was assumed that each grid would have a relatively low stiffness of 2 MN/m, thus giving a 'total' stiffness of 3 × 2 = 6 MN/m, as previously used for a single grid.

The sequence of analysis was the same as the previous chapter. First the in-situ stresses were specified (again based on a unit weight of 17 kN/m^3 and a K_0 value of 0.5). σ_s then reduced from the nominal vertical stress at the base of the embankment to mimic the loss of support from the subsoil.

Table 4.1 Material parameters for geogrid

Young's modulus (MN/m^2)	Poisson's ratio	Cross-section area (m^2/m)
6000 (or 12,000 or 2000 with three layers of geogrid)	0.0	0.001

Table 4.2 Summary of analyses reported in this chapter

	h (m)	s (m)	J (MN/m)	δ_i	subplot
Effect of h	1	2.5	6	0	(a)
	3.5	2.5	6	0	(b)
	10	2.5	6	0	(c)
Effect of s	3.5	2	6	0	(d)
	3.5	3.5	6	0	(e)
Effect of geogrid: J	3.5	2.5	12	0	(f)
Effect of geogrid: δ_i	3.5	2.5	6	20	(g)
Effect of geogrid: N	3.5	2.5	3 × 2	0	(h)
Effect of geogrid: N and δ_i	3.5	2.5	3 × 2	20	(i)

4.3.2 For Three-Dimensional Condition

The analyses reported here are essentially an extension of plane strain case, with the addition of a single or three layers of reinforcement near the base of the embankment. The FE mesh is a simple cuboid $s/2$ square in plan and h high. This represents the embankment corresponding to one quarter of a pile cap (Fig. 4.3). The embankment fill was modelled as a dry linear elastic-perfectly plastic granular soil.

The base of the embankment mesh is supported rigidly by the pile cap, and elsewhere by a uniform 'subsoil stress'. The pile and subsoil are modelled via these boundary conditions, again corresponding to previous chapters. The subsoil stress is used to control the analysis, starting at the nominal overburden value (γh), and reducing to stimulate arching in the embankment and sag of the reinforcement. With

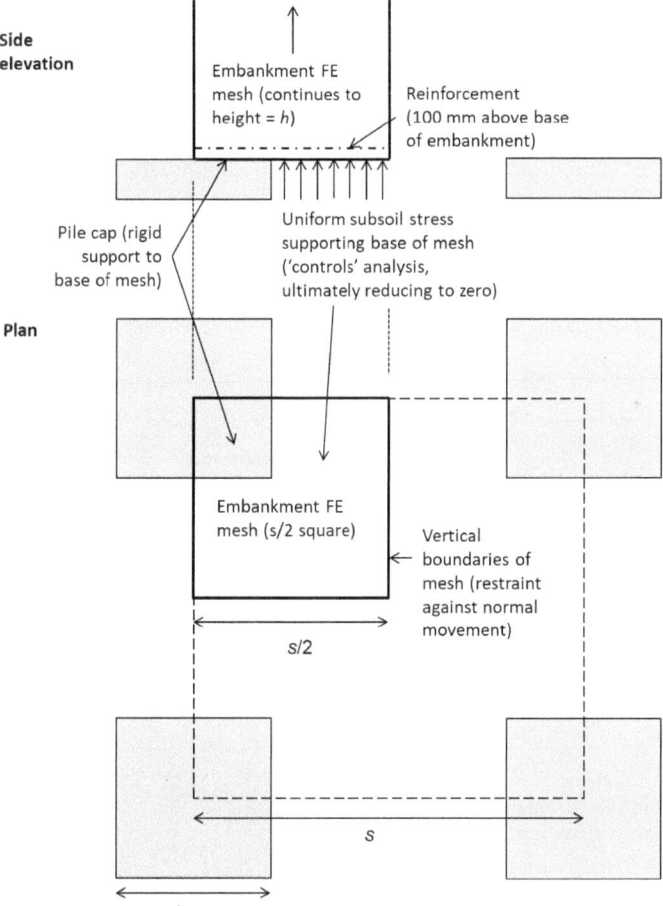

Fig. 4.3 Geometry and mesh boundaries for three-dimensional condition

the inclusion of reinforcement, it was possible to effectively reduce the subsoil stress to zero (or in fact a small value corresponding to the 100 mm thickness of embankment material below the reinforcement, Fig. 4.3). This loss of subsoil support is broadly analogous to consolidation of the subsoil with time after construction.

The FE analyses reported in Chap. 3 only considered arching in the embankment (without reinforcement). The results showed reasonable correspondence with predictions from the Hewlett and Randolph (1988) method, which were about 40% higher than FE results, i.e. a 'conservative' prediction. The importance of the mechanism of failure 'at the pile cap' (punching into the base of the embankment) was also confirmed, noting that this is not considered in the Marston approach in BS 8006.

Table 4.3 shows the significant range of geometries considered for three-dimensional condition in this chapter. Generally $a = 1.0$ m, but 0.5 m is also considered. The 'spacing ratio' (s/a) increases from 2.0 to 3.5, corresponding to area ratio reducing from 25 to 8%. The height ratio $h/(s - a)$ has minimum value 2.0 to allow a 'full' arch to form (Fig. 4b; Zhuang et al., 2012), whilst the maximum height is 10.0 m.

The 'standard' size of 20-noded, reduced-integration, quadratic brick solid elements ('C3D20R') was expressed as a multiple of the (smallest) length scale, l (Table 4.3), as shown in Table 4.4. The vertical dimension of elements was twice the horizontal, and elements in the 'upper' portion of the embankment (height $> 2(s - a)$) were twice the dimension of elements in the 'lower' portion (where height $< 2(s - a)$ and the main effect of arching occurred). The smallest elements were 20.8 mm (horizontal) $\times$ 41.7 mm (vertical) in the lower part of the embankment when $s = 1.25$ m. The largest elements were 100 mm (horizontal) $\times$ 200 mm (vertical) in the upper part of the embankment when $s = 3.5$ m.

Table 4.3 Geometries for FE analyses

Pile cap a	C–C spacing s	Clear spacing $s - a$	Spacing ratio s/a	Embankment height h, m		Height ratio $h/(s - a)$		FE mesh length scale l
m	m	m		min	max	min	max	mm
1.0	2.0	1.0	2.0	2.0	10.0	2.0	10.0	33.3
0.5	1.25	0.75	2.5	1.5	6.5	2.0	8.7	20.8
1.0	2.5	1.5	2.5	1.0	10.0	0.7	6.7	41.7
1.0	3.5	2.5	3.5	3.5	10.0	1.4	4.0	50.0

Table 4.4 Brick element dimensions in terms of length scale l (defined in Table 4.3)

	Height in embankment	Element dimension	
		Horizontal	Vertical
'Upper'	$> 2 (s - a)$	$2l$	$4l$
'Lower'	$< 2 (s - a)$	l	$2l$

The reinforcement was modelled using 4—noded, full integration, 3-dimensional membrane elements ('M3D4') elements, which have tensile stiffness, J (MN/m), without bending stiffness. The interface friction angle (δ_i) between the granular material (embankment) and geogrid was $0°$, $20°$ or $30°$. The corresponding geogrid stiffness for nominal 1 mm thickness was 6 MN/m, 12 MN/m, or 2 MN/m with three layers of geogrid. Two separate 'upper' and 'lower' orthogonal layers were not modelled, and hence the result corresponds to an 'average' which would be more than an upper layer and less than a lower layer (Love & Milligan, 2003).

Meshes contained approximately 25–70 thousand elements. The sensitivity to mesh size was checked for each value of s using the intermediate embankment height (and low reinforcement stiffness where this was varied). l was reduced by a factor (3/4), i.e. increasing the number of elements by a factor $(4/3)^3 = 2.4$. This caused the maximum tension in the reinforcement to increase by between 1 and 4%, whilst the maximum sag reduced between 1 and 3%, values which were considered suitably small.

4.4 Results

The results for the plane strain and three-dimensional conditions including behaviour of reinforced piled embankment, settlement at the subsoil and surface of the reinforced piled embankment, and behaviour of geogrid in the piled reinforced embankment will be presented and discussed in the following section. Meanwhile, the predictions for the performance of geogrid by Hewlett and Randolph, Marston and BS 8006 approaches will be presented and compared with results from three-dimensional Finite Element analyses. Furthermore, a suggested modification to the BS 8006 formula for prediction of reinforcement sag will be proposed in this section.

4.4.1 For Plane Strain Condition

4.4.1.1 Behaviour of Reinforced Piled Embankment

Figure 4.4 shows the maximum displacement at the midpoint between pile caps (δ_s), which increases with a reduction in the stress on the subsoil (σ_s). There are three lines:

- 'Ground Reaction Curve' (GRC)
- 'Embankment with geogrid'
- 'GRC + effect of geogrid' (a theoretical comparison line)

The first part of the GRC line comes from the analyses reported in Chap. 3, for an unreinforced piled embankment. These results were previously presented in Fig. 3. 4 in a normalised form. The value of δ_s at the point of maximum arching has been 'extrapolated' to large δ_s.

The 'embankment with geogrid' line shows results from the analyses summarised in Table 4.2 as separate sub-plots. Note that 'geogrid' could equally refer to geosynthetic reinforcement. The 'GRC + effect of geogrid' line, has been derived from the ground reaction curve (GRC) combined with Eq. (2.30).

If the GRC data from Chap. 3 is δ_s (GRC), and $W_T(\delta_g = \delta_s)$ is the reduction in stress on the subsoil due to the vertical stress carried by the geogrid for a sag equal to the subsoil settlement, then

$$\sigma_s(\text{GRC} + \text{effect of geogrid}) = \sigma_s(\text{GRC}) - W_T(\delta_g = \delta_s).$$

Here, the length of span (l) is assumed to be ($s - a/2$)—the origin of this value will be explained later. This expression can then be evaluated for any value of σ_s where GRC data is available, and is hence plotted in Fig. 4.4 as a 'theoretical comparison' line.

In general the response from an analysis of the 'Embankment with geogrid' is similar to the comparison ('GRC + effect of geogrid') line. However, in some cases [subplots (a) and (e)] the embankment with geogrid analysis performs 'better' than anticipated based on the comparison line (i.e. σ_s is less than predicted for a given δ_s). Generally the data points on the 'Embankment with geogrid' line are so densely positioned (corresponding to increments in the analysis) that they cannot be clearly identified. However, the extrapolated 'GRC' and 'GRC + effect of geogrid' are clearly identified and thus it can be established that the 'Embankment with geogrid' line is the remaining line.

For the GRC, the stress at the base of the embankment (σ_s) never reaches zero. However, if the geogrid carries the remaining stress at the point of maximum arching, σ_s can reach to zero, and approaches this value at the end of the analyses. However, significant sag of the geogrid is required for this to happen.

Subplots (a), (b) and (c) show the effect of increasing embankment height (h). The subsoil settlement at the midpoint between piles (δ_s) when $\sigma_s \approx 0$, increases with increasing embankment height. This is because σ_s at the point of maximum arching for the GRC increases with the height of the embankment (see Fig. 3.4b in Chap. 3), and hence the geogrid has to carry more load and deforms more.

Subplots (d), (b) and (e) show the effect of increasing centre-to-centre pile cap spacing (s). The largest settlement of the subsoil is observed at the largest s, corresponding to very strong dependency on l in Eq. (2.30). For subplot (f), comparing with subplot (b) there is relatively little change, corresponding to relatively limited dependency on geogrid stiffness (J) in Eq. (2.30). Subplot (g) shows the effect of increased interface friction angle (δ_i) between the geogrid and embankment soil. Compared to subplot (b) the data show the settlement of the subsoil when $\sigma_s \approx 0$ is slightly reduced, but there is not a major impact.

Subplot (h) considers three layers of geogrid ($J = 2$ MN/m) with a frictionless interface. This causes the point of maximum arching in the analysis for the full embankment with geogrid to have slightly higher σ_s than the GRC. This can be attributed to the presence of the three layers of geogrid with frictionless interfaces, which weakens the mass properties of the soil near the base of the embankment. The

Fig. 4.4 Variation of subsoil settlement and stress for reinforced piled embankments; **a** $h = 1$ m, $s = 2.5$ m, $J = 6$ MN/m, $\delta_i = 0$, **b** $h = 3.5$ m, $s = 2.5$ m, $J = 6$ MN/m, $\delta_i = 0$, **c** $h = 10$ m, $s = 2.5$ m, $J = 6$ MN/m, $\delta_i = 0$, **d** $h = 3.5$ m, $s = 2$ m, $J = 6$ MN/m, $\delta_i = 0$, **e** $h = 3.5$ m, $s = 3.5$ m, $J = 6$ MN/m, $\delta_i = 0$, **f** $h = 3.5$ m, $s = 2.5$ m, $J = 12$ MN/m, $\delta_i = 0$, **g** $h = 3.5$ m, $s = 2.5$ m, $J = 6$ MN/m, $\delta_i = 20$ (interface friction angle), **h** $h = 3.5$ m, $s = 2.5$ m, $J = 3 \times 2$ MN/m, three layers of geogrid with $\delta_i = 0$ (interface friction angle), **i** $h = 3.5$ m, $s = 2.5$ m, $J = 3 \times 2$ MN/m, three layers of geogrid with $\delta_i = 20$ (interface friction angle)

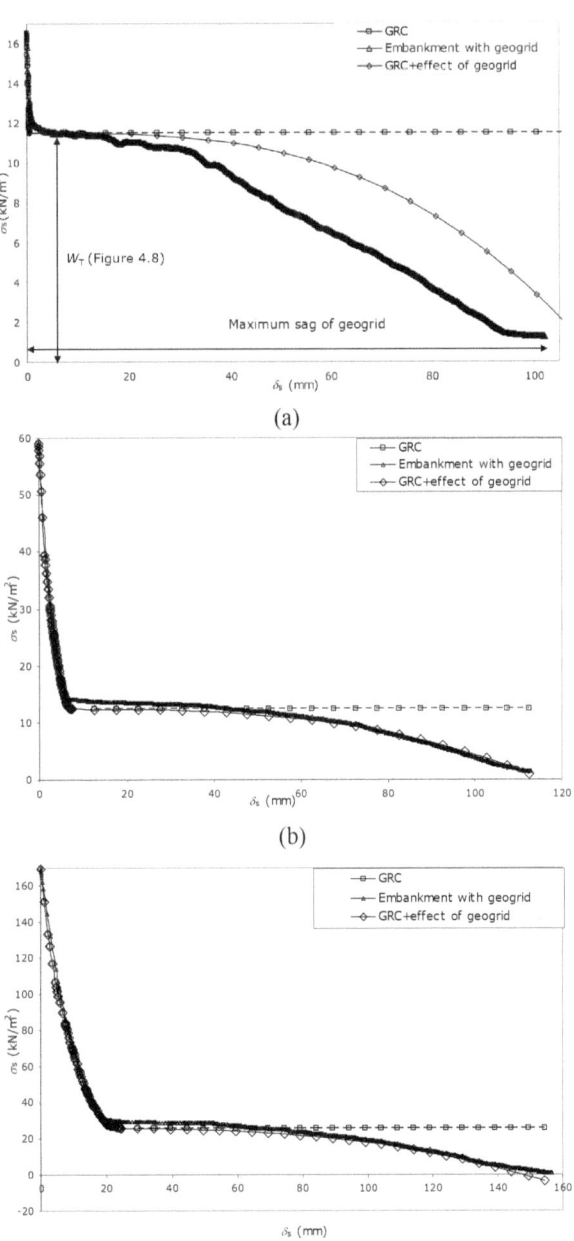

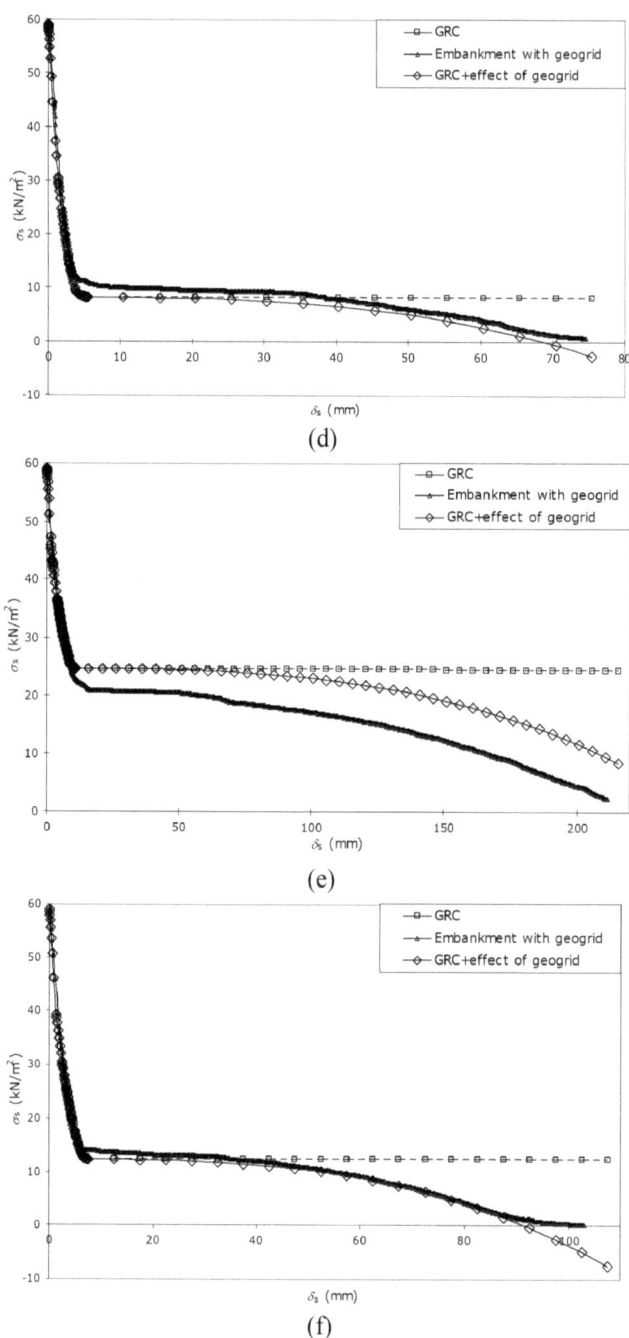

Fig. 4.4 (continued)

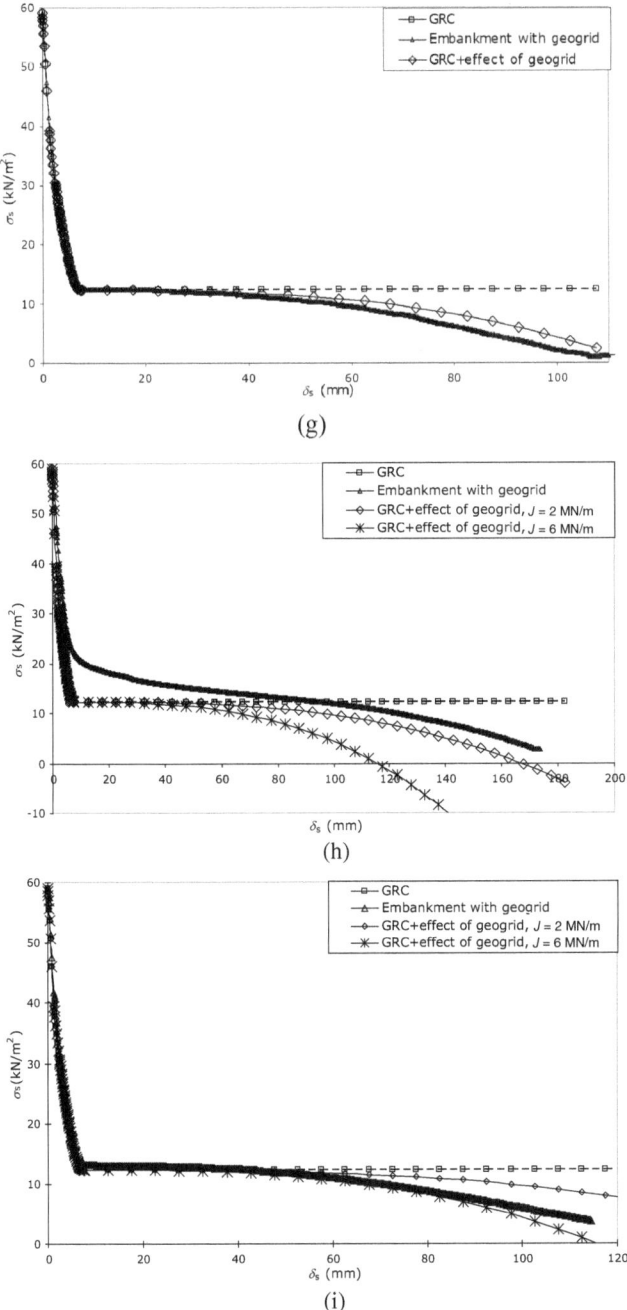

Fig. 4.4 (continued)

results at larger displacement are closer to a comparison line based on $J = 2$ MN/m (for a single geogrid). This implies that the geogrids are less effective than a single geogrid with 3 times the stiffness. The tension in the geogrids will be considered further below.

Subplot (i) shows the result for three layers of geogrid with $J = 2$ MN/m, where the interface friction angle between the geogrid and the surrounding soil (δ_i) is 20°. This improves comparison with the GRC at the point of maximum arching, and the data are close to the comparison line for $J = 6$ MN/m, which reflecting the total stiffness of the three grids.

4.4.1.2 Settlement at the Subsoil and Surface of the Reinforced Piled Embankment

Figure 4.5a shows the maximum value of subsoil settlement at the midpoint between piles (see Sect. 2.1, Fig. 2.1) required to reach ultimate conditions: $\sigma_{s'ult} \approx 0$. The value has been normalised by the clear gap between pile caps ($s - a$) so that it is analogous to δ^* see Sect. 2.3, Fig. 2.13). Variation with (h/s) is shown [from subplots (a) to (c)] throughout Fig. 4.5.

The magnitude of δ^* is between 6 and 11% when $\sigma_{s'ult} \approx 0$ and the normalised displacement to reach this point increases with h/s. These values are considerably larger than the equivalent data for the point of maximum arching, which were approximately between 0 and 1.6% (see Chap. 3, Fig. 3.4a). This finding can be explained by the significant geogrid sag required to carry the remaining subsoil stress.

As presented in Fig. 4.5b, the ratio of settlement at the top of the embankment at the midpoint between piles (δ_{em} see Sect. 2.1, Fig. 2.1) to the equivalent value in the subsoil (δ_s) at the point where $\sigma_{s'ult} \approx 0$: (δ_{em}/δ_s) is in the range 0.4–0.9. This is similar to the result reported in Chap. 3 (see Chap. 3, Fig. 3.4b) for the point of maximum arching on the GRC, and only approaches 1.0 when the embankment is very low and there is no arching. For higher embankments settlement at the surface of the embankment is less than at the subsoil as would be expected.

Figure 4.5c shows the ratio of δ_{em} to the equivalent value at the centreline above the pile cap δ_{ec} (see Sect. 2.1, Fig. 2.1) at the point where $\sigma_{s,ult} \approx 0$: (δ_{em}/δ_{ec})$_{ult}$, showing variation with (h/s). This is a measure of differential settlement at the surface of the embankment, which shows similar behaviour to the point of maximum arching for the GRC (see Chap. 3, Fig. 3.4c). For (h/s) > 1.5 the ratio is 1.0, which indicates there is no differential settlement at the top of the embankment. However, the differential settlement increases dramatically for the lowest embankment.

4.4.1.3 Behaviour of Geogrid in the Piled Reinforced Embankment

Figure 4.6 shows the amount of vertical load which is carried by the geogrid (w), illustrating variation with the maximum sag of the geogrid [subplot (a)], and the

Fig. 4.5 Ultimate ($\sigma_{s,ult} \approx 0$) settlement at the subsoil and surface of the reinforced piled embankment; **a** ultimate settlement of the subsoil at the midpoint between piles ($\delta_{s,ult}$) normalised by the clear gap between pile caps ($s - a$), **b** ratio of the settlement at the top of the embankment at the midpoint between piles (δ_{em}) to the equivalent value at the subsoil ($\delta_{s,ult}$), **c** ratio of the settlement at the top of the embankment at the midpoint between piles (δ_{em}) to the equivalent value at the centreline above the pile cap (δ_{ec})

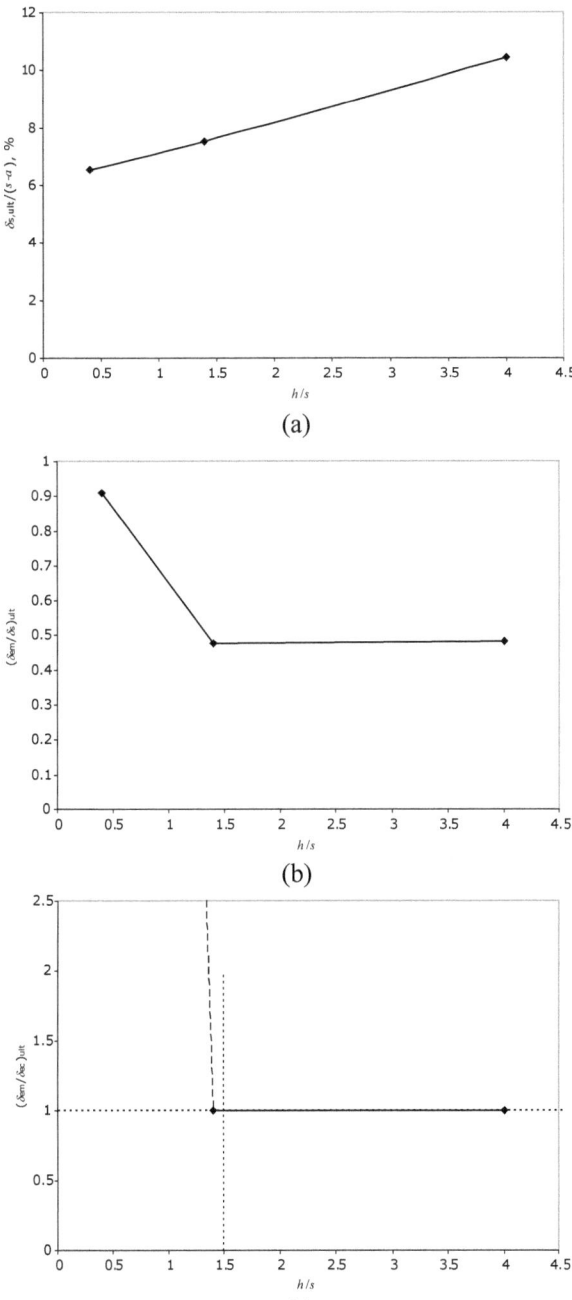

tension which this load generates in the geogrid [subplot (b)] respectively. Figure 4.4 shows how these values were established from the plots.

Theoretical comparison lines were derived by combining Eqs. (2.26), (2.27) and (2.28). Substituting Eq. (2.27) into Eq. (2.28) leads to:

$$T_{rp} = \tfrac{8}{3} J \left(\tfrac{\delta_g}{l} \right)^2 \tag{4.7}$$

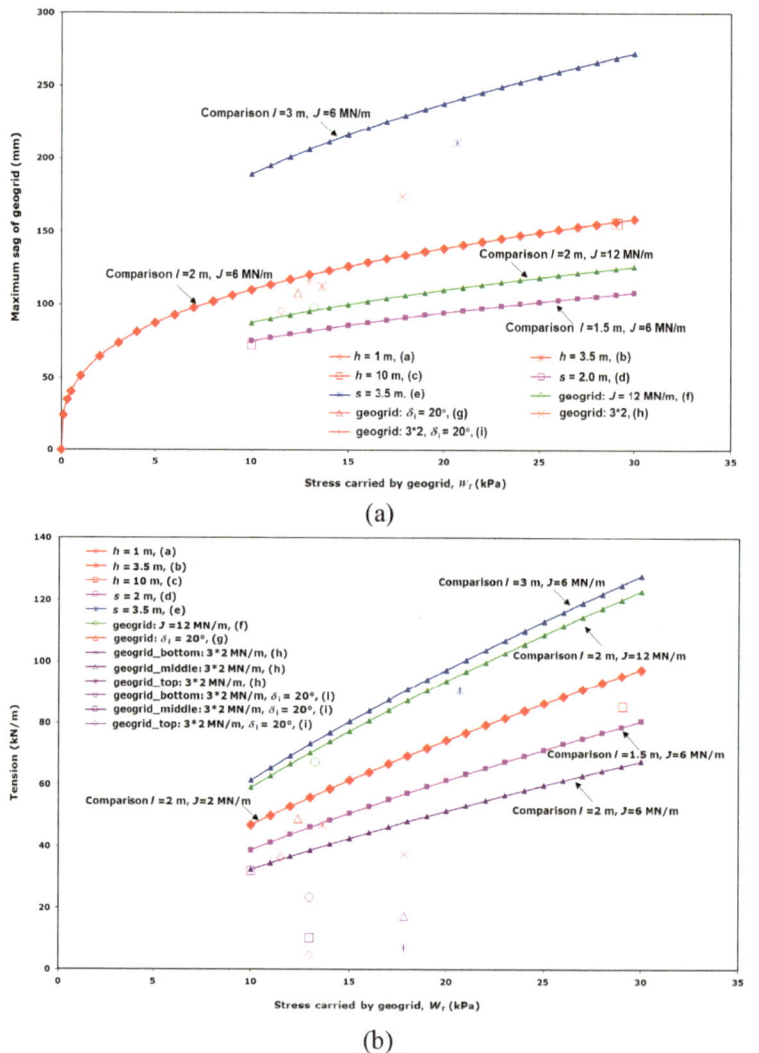

Fig. 4.6 Maximum displacement and tension of geogrid generated by vertical stress carried by the geogrid (W_T). Specific colours associate results with comparison lines; **a** maximum sag of the geogrid, **b** tension in the geogrid

Re-arranging this equation:

$$\sqrt{\frac{3T_{rp}}{8J}} = \frac{\delta_g}{l} \qquad\qquad (4.8)$$

Re-arranging Eq. (2.26):

$$\frac{lW_T}{8T_{rp}} = \frac{\delta_g}{l} \qquad\qquad (4.9)$$

Combing Eqs. (4.8) and (4.9),

$$\sqrt{\frac{3T_{rp}}{8J}} = \frac{lW_T}{8T_{rp}} \qquad\qquad (4.10)$$

Thus,

$$T_{rp} = \left(\tfrac{1}{24}JW_T^2 l^2\right)^{(1/3)} \qquad\qquad (4.11)$$

Equation (2.29) can be re-arranged to give:

$$\delta_g = l\left[\tfrac{3}{64}\left(\tfrac{W_T l}{J}\right)\right]^{(1/3)} \qquad\qquad (4.12)$$

Equations (4.11) and (4.12) were used to derive the comparison lines in Fig. 4.6b and (a) respectively with $l = (s - a/2)$. This value was chosen as an average of the unsupported span $(s - a)$ and the centre-to-centre spacing s.

Figure 4.6 shows one data point for each of the analyses summarised in Table 4.2. This corresponds to the ultimate point in the analysis, when $\sigma_{s,ult} \approx 0$ and sag of the geogrid has reached its maximum value. The corresponding stress carried by the geogrid W_T, is taken as the value of σ_s at the point of maximum arching.

Subplot (a) shows stress on the geogrid and the corresponding maximum sag. A total of 4 comparison lines (Eq. 4.12) are shown, corresponding to variation of s and J. The comparison line for $l = 2$ m ($s= 2.5$ m) and $J = 6$ MN/m corresponds to cases (a–c) and (g–i) in Table 4.2. Hence the line and corresponding data points are red. The data shows reasonable agreement except for analysis (h)—3 geogrids with frictionless interface with the soil. Since the sum of stiffness for the 3 grids is 6 MN/m, it can be compared with this line. However, the displacement is somewhat larger than expected, which as previously noted reflects the reduced effectiveness of the upper grids in carrying load.

For $l = 1.5$ m ($s = 2.0$ m) with $J = 6$ MN/m (pink), comparison with case (d) is good. For $l = 3$ m ($s = 3.5$ m) with $J = 6$ MN/m (blue), the correct trend of behaviour can be observed, but the comparison line somewhat overestimates the displacement compared to case (e). For $l = 2.0$ m ($s = 2.5$ m) with $J = 12$ MN/m (green), comparison is good with case (f). In general, the results show good comparison with Eq. (4.12), confirming that the maximum sag is considerable more affected by the span l than the stiffness J.

Subplot (b) shows that the tension in the reinforcement increases with an increase of the vertical stress carried by the geogrid W_T. For cases (a–g) the comparison lines are matched with data points from the analyses using the same colours as in Subplot (a). It was observed that tension in the geogrid was approximately constant across the width, including over the pile cap. All the check lines give a slightly conservative estimate of tension compared to the data, but agreement is reasonable.

For the cases with three layers a separate data point is shown for each layer of geogrid, and a new comparison line is shown based on $J = 2$ MN/m (purple), the stiffness of each geogrid rather than the combined total for all 3 ($J = 6$ MN/m). It can be seen that the upper two grids carry relatively little tension compared to the bottom layer, presumably implying less sag Eq. (4.7) and less effective performance. In general the results show good comparison with Eq. 4.11, confirming that tension is most sensitive to the stress carried by the geogrid and length of the span.

4.4.2 For Three-Dimensional Condition

4.4.2.1 Behaviour of Reinforced Piled Embankment

The results in Fig. 4.7 are presented in the same format as previous sections:

- 'GRC' line
- 'Embankment with geogrid' line (data derived in this chapter)
- 'GRC + effect of geogrid' (theoretical comparison)

The theoretical comparison line is again based on Eq. (2.30), but here using a span ($l = \sqrt{2}\,(s - a)$), based on the diagonal clear span between pile caps, and accepting that there is some idealisation since the equation is plane strain. The subsoil settlement plotted (δ_s) is the maximum on the diagonal between pile caps (A, Fig. 3.2). The analysis of the 'Embankment with geogrid' is in general similar to the comparison line derived from the GRC combined with Eq. (2.30), although agreement is not quite as good as in the plane strain analyses.

For subplots (a), (b) and (c), the subsoil settlement at the midpoint of the diagonal between piles δ_s when $\sigma_s = 0$ increases with the height of the embankment, since the geogrid has to carry more load and deforms more. Subplots (d), (b) and (e) show the maximum settlement (δ_s) increases with the centre-to-centre pile cap spacing (s).

Again for subplot (f), comparing with subplot (b), δ_s when $\sigma_s = 0$ reduces slightly, reflecting the effect of J. Subplot (g) indicates that δ_s for $\sigma_s = 0$ reduces slightly as the interface friction angle between the geogrid and subsoil increases, but as shown in plane strain there is not a major impact.

Subplot (h) shows the effect of three layers of low stiffness ($J = 2$ MN/m) geogrid with a frictionless interface. As shown in plane strain, this causes the point of maximum arching in the analysis for the embankment with geogrid to have slightly

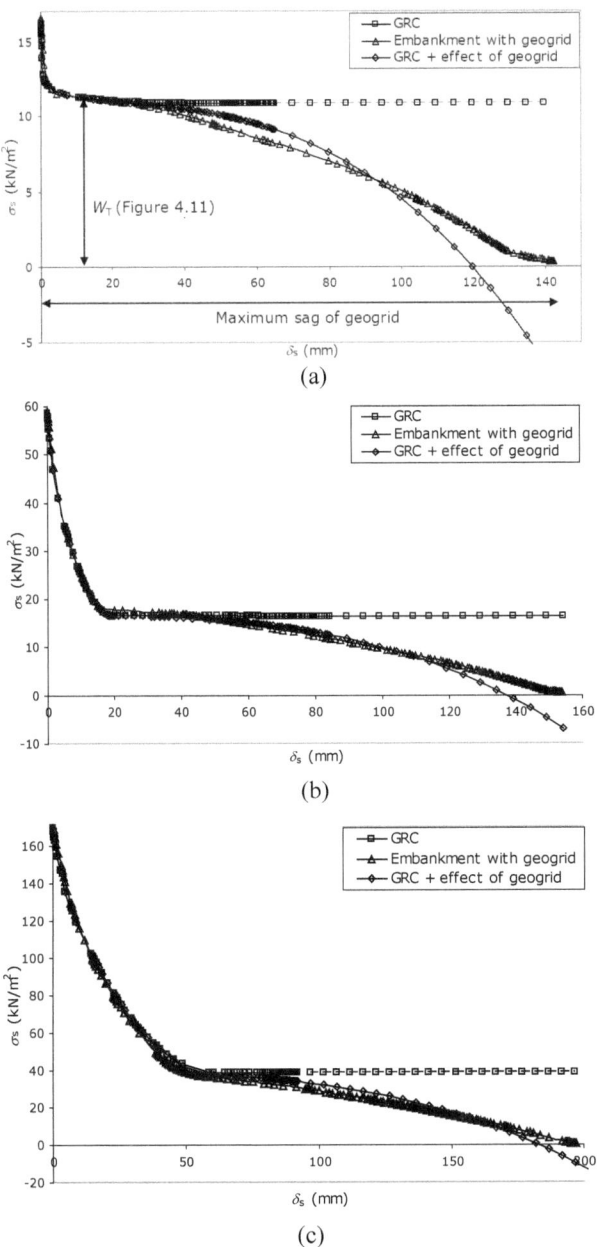

Fig. 4.7 Variation of subsoil settlement and stress for reinforced piled embankments; **a** $h = 1$ m, $s = 2.5$ m, $J = 6$ MN/m, $\delta_i = 0$, **b** $h = 3.5$ m, $s = 2.5$ m, $J = 6$ MN/m, $\delta_i = 0$, **c** $h = 10$ m, $s = 2.5$ m, $J = 6$ MN/m, $\delta_i = 0$, **d** $h = 3.5$ m, $s = 2$ m, $J = 6$ MN/m, $\delta_i = 0$, **e** $h = 3.5$ m, $s = 2.5$ m, $J = 6$ MN/m, $\delta_i = 0$, **f** $h = 3.5$ m, $s = 2.5$ m, $J = 12$ MN/m, $\delta_i = 0$, **g** $h = 3.5$ m, $s = 2.5$ m, $J = 6$ MN/m, $\delta_i = 20$ (interface friction angle), **h** $h = 3.5$ m, $s = 2.5$ m, $J = 3 \times 2$ MN/m, three layers of geogrid with $\delta_i = 0$ (interface friction angle), **i** $h = 3.5$ m, $s = 2.5$ m, $J = 3 \times 2$ MN/m, three layers of geogrid with $\delta_i = 20$ (interface friction angle)

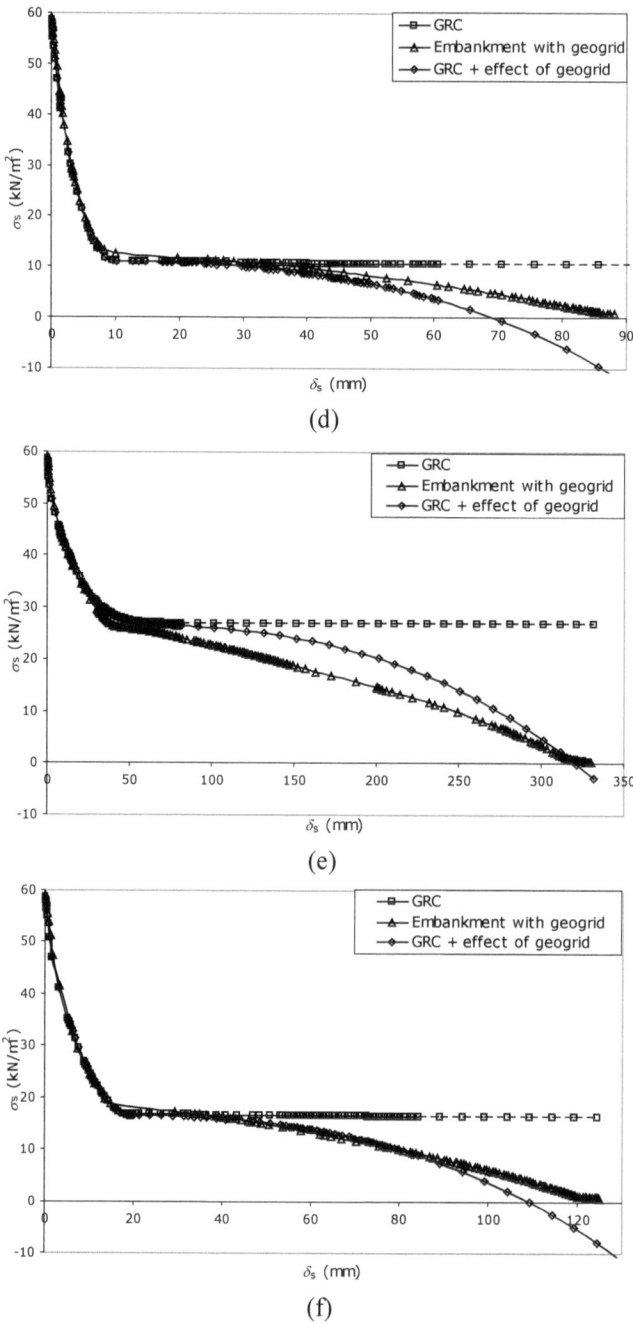

Fig. 4.7 (continued)

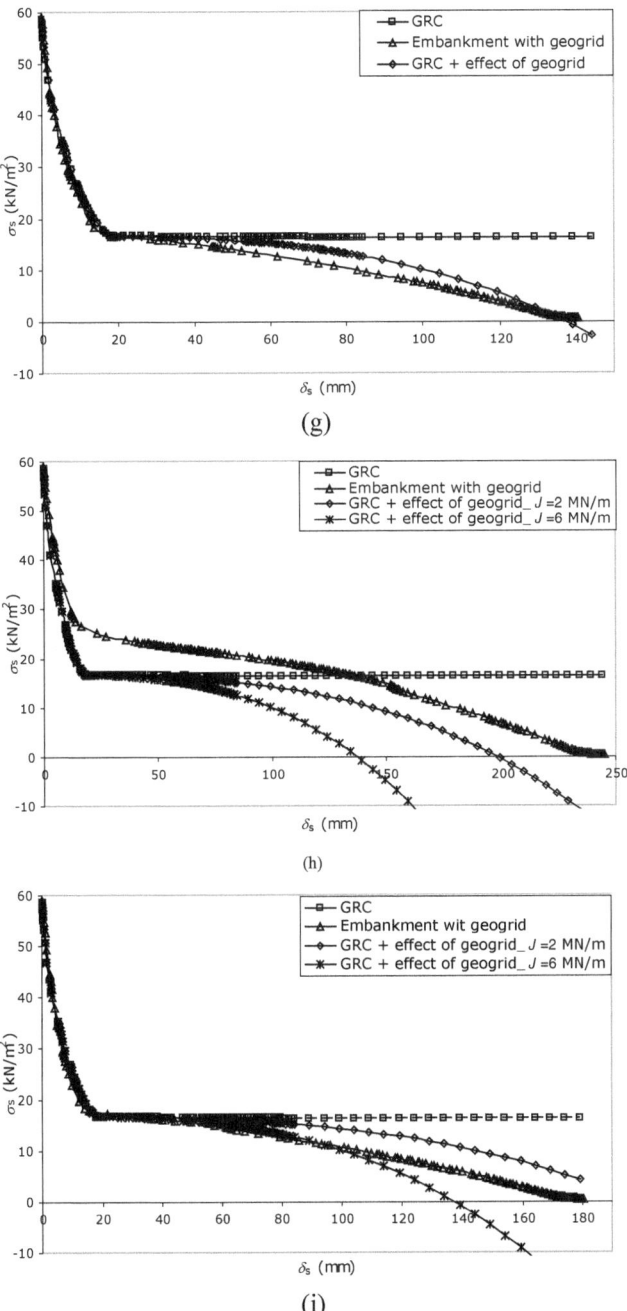

(g)

(h)

(i)

Fig. 4.7 (continued)

higher σ_s than the GRC. This was attributed to the damaging effect of the friction-less interfaces created within the soil. Furthermore, the data ultimately show better correspondence with Eq. (2.30) based on $J = 2$ MN/m rather than the total stiffness of 6 MN/m.

Subplot (i) considers three layers of geogrid ($J = 2$ MN/m) and the interface friction angle between the geogrid and embankment material is 20°. This improves comparison with the GRC at the point of maximum arching, and the data now lies between the comparison line for $J = 2$ and 6 MN/m.

4.4.2.2 Settlement at the Subsoil and Surface of the Reinforced Piled Embankment

Figure 4.8a shows the maximum value of subsoil settlement at the midpoint of the diagonal between piles (A, Fig. 3.2) required to reach ultimate conditions: $\sigma_{s,ult} \approx 0$. The value has been normalised by the clear gap between pile caps ($s - a$) so that it is analogous to δ^*. Variation with (h/s) is shown [for analyses (a–c)]. The absolute magnitude of $\delta_{s,ult}/(s - a)$ is between 10 and 13% when $\sigma_s \approx 0$. The normalised displacement to reach this point increases slightly with h/s. These values are some-what larger than the equivalent data in the plane strain analyses (approximately between 6 and 11%). Because the reinforcement span is for the 3D situation, larger displacement is required to reach the point where $\sigma_s \approx 0$.

Subplot (b) shows the ratio of settlement at the top of the embankment at the midpoint of the diagonal between piles (δ_{em}, see plane strain graph Fig. 2.1, in Sect. 2.1) to the equivalent value in the subsoil (δ_s) at the point where $\sigma_s \approx 0$: (δ_{em}/δ_s)$_{ult}$. Values are in the range 0.6–0.8. This is similar to the plane strain situation, where values were in the range 0.4–0.9. The ratio again tends to 1.0 as (h/s) tends to 0.

Subplot (c) shows the ratio of the settlement at the top of the embankment at the midpoint of the diagonal between piles (δ_{em}) to the equivalent value at the centre above the pile cap δ_{ec}, see plane strain graph Fig. 2.1, in Sect. 2.1) at the point where $\sigma_s \approx 0$: (δ_{em}/δ_{ec})$_{ult}$, showing variation with (h/s). This is a measure of differential settlement at the surface of the embankment, which shows similar behaviour to the plane strain results, and 3D results for an embankment without geogrid.

4.4.2.3 Results for Reinforcement Tension

In the previous section (the equivalent plane strain analyses), equations were derived relating both the maximum sag and tension in the geogrid to the component of vertical stress carried by the geogrid (W_T). This is the vertical stress at the base of the embankment which 'remains' at the point of maximum arching and is carried by the geogrid when the subsoil stress (σ_s) reaches zero in the analysis. As in previous section the plots show one data point for each of the analyses. This corresponds to the ultimate point in the analysis, when $\sigma_{s,ult} \approx 0$ and sag of the geogrid has reached

Fig. 4.8 Ultimate ($\sigma_{s,ult} \approx$ 0) settlement at the subsoil and surface of the reinforced piled embankment; **a** ultimate settlement of the subsoil at the midpoint of the diagonal between piles ($\delta_{s,ult}$) normalised by the clear gap between pile caps ($s - a$), **b** ratio of the settlement at the top of the embankment at the midpoint of the diagonal between piles (δ_{em}) to the equivalent value at the subsoil ($\delta_{s,ult}$), **c** ratio of the settlement at the top of the embankment at the midpoint of the diagonal between piles (δ_{em}) to the equivalent value at the centre above the pile cap (δ_{ec}) at ultimate conditions

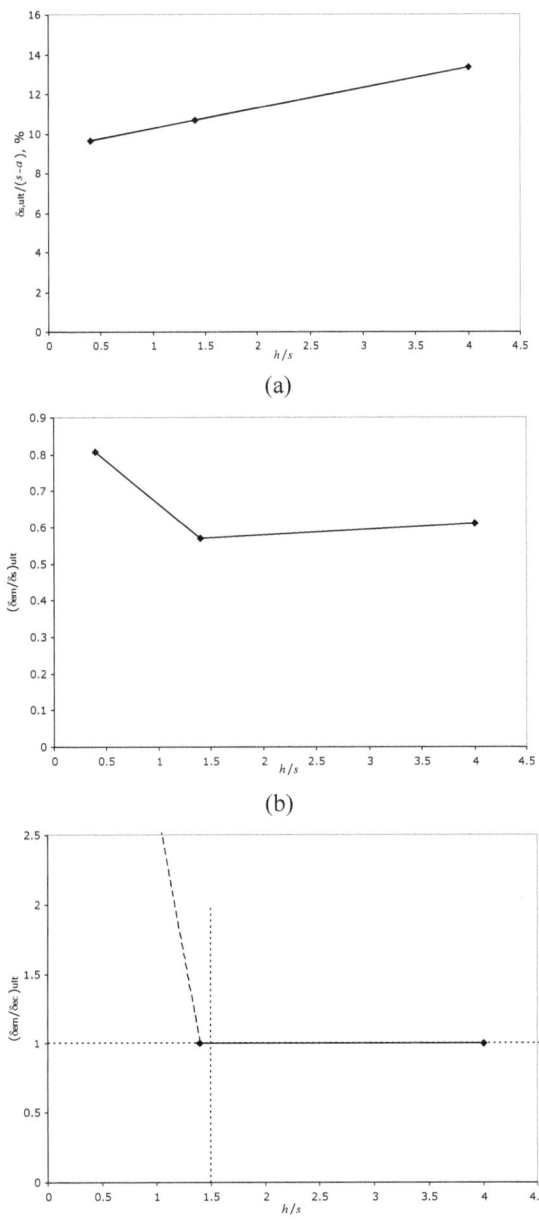

its maximum value. The value of w in each case was derived from the GRC at the point of maximum arching.

Figure 4.9a shows results for the maximum sag in the geogrid, which occurred at the midpoint of a diagonal between pile caps (point C, Fig. 3.2). The four comparison lines were generated using Eq. (4.12) with $l = \sqrt{2}\,(s - a)$. The factor $\sqrt{2}$ is associated with the change from plane strain to 3D geometry. The use of colours to associate specific data points with each of the comparison lines is the same as the plane strain case. As in that case the data shows reasonable agreement except for analysis (h)—3 geogrids with frictionless interface with the soil. Since the sum of stiffness for the 3 grids is 6 MN/m, it can be compared with this line. However, the displacement is again larger than expected, which as previously noted probably reflects the reduced effectiveness of the upper grids in carrying load.

Figure 4.9b shows the tension in the reinforcement at the midpoint of a diagonal between piles, which increases with an increase of the remaining vertical load carried by the geogrid. Comparison lines were generated using Eq. (4.11) with $l = \sqrt{2}\,(s - a)$. However, the comparison lines overestimate the analysis results.

In fact, the maximum tension in the geogrid occurred at the corner of a pile cap, and was about 2–3 times higher than the value at the midpoint of the diagonal. Figure 4.10 shows typical contours of this tension in both the orthogonal directions. It has been noted by (Russell & Pierpoint, 1997) that tension will not be maximum at the centre of the span, and this has been confirmed here in 3D (although tension across the span and pile cap was virtually constant in plane strain). Figure 4.9c shows the maximum tension for the various analyses. The tension T_{rp} predicted by Eq. 4.11, is modified by the factor $(s + a)/2a$, and referred to as T^*:

$$T_{rp}^* = T_{rp}\,\frac{(s+a)}{2a} \tag{4.13}$$

This factor was proposed by Love and Milligan (2003) for the increased tension in geogrid between piles where the load is distributed evenly in both orthogonal directions, corresponding to the situation in the analyses. The comparison lines are then generated by Eq. (4.13) with $l = (s - a/2)$ in Eq. 4.11 to derive T_{rp}, corresponding to plane strain case, since a span directly between pile caps rather than a diagonal is considered. The lines in Fig. 4.9c are therefore the lines in Fig. 4.9b multiplied by $(s + a)/2a$.

The data now generally show good agreement with the comparison lines. For the cases with three reinforcement layers a separate data point is shown for each layer of geogrid, and a new comparison line is shown based on $J = 2$ MN/m (purple), the stiffness of each geogrid rather than the combined total for all 3 ($J = 6$ MN/m). It can be seen that like in plane strain the upper two grids carry relatively little tension compared to the bottom layer, implying less effective performance. This finding has been proposed by Jenner et al. (1998) for a monitored field case. They stated that larger strains were recorded in the lower grid than the upper grid as anticipated in the design.

Figure 4.11 compares the predictions of T_{rp} from Eq. 4.4 with the maximum tension from the FE analyses, the Marston prediction, and Hewlett and Randolph

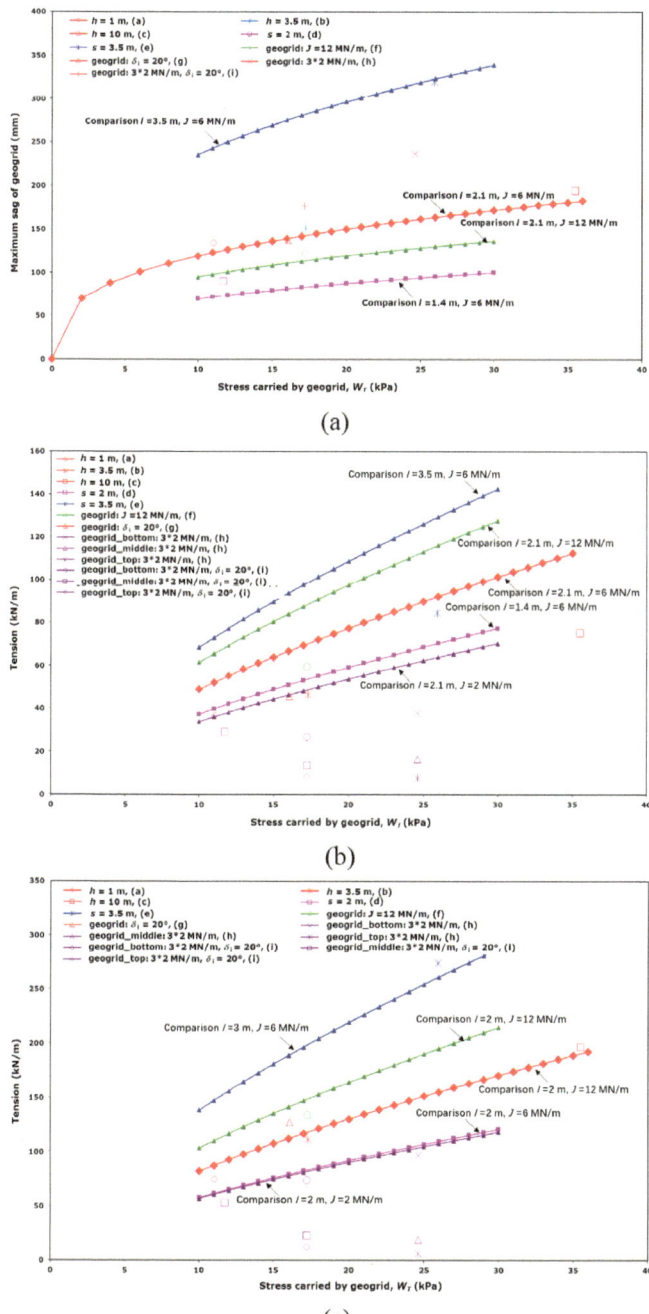

Fig. 4.9 Maximum displacement and tension of geogrid generated by vertical stress carried by the geogrid (W_T). Specific colours associate results with comparison lines; **a** maximum sag of the geogrid, **b** tension in the geogrid at midpoint of diagonal between piles (A, see Fig. 3.2), **c** tension in the geogrid at the corner of the pile cap (C, see Fig. 3.2)

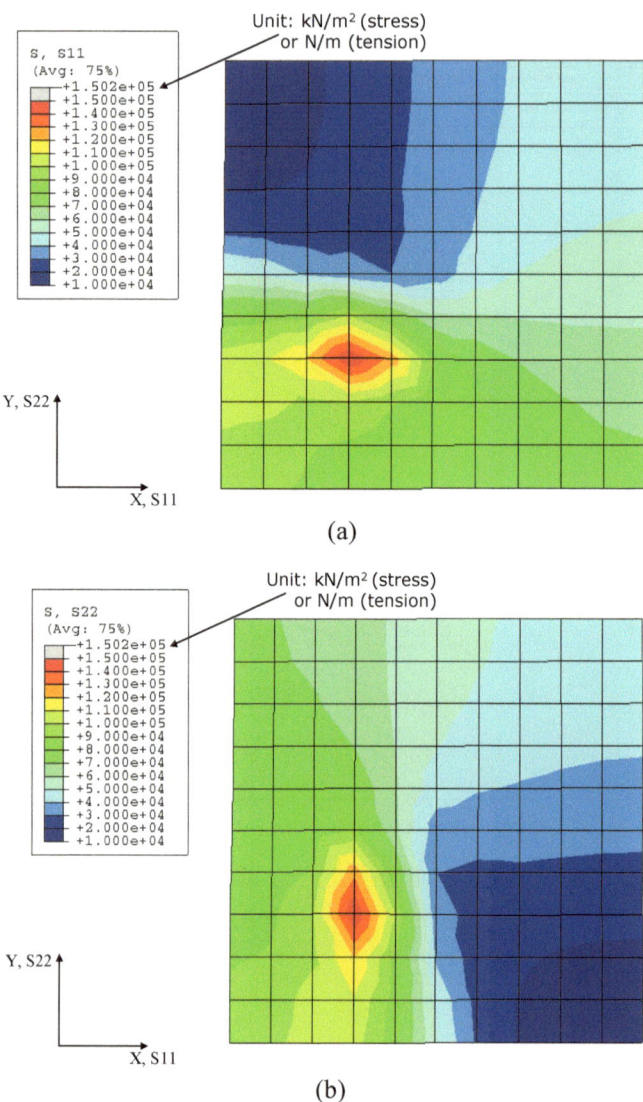

Fig. 4.10 Tension distribution of geogrid at the maximum sag; **a** X direction, **b** Y direction

approaches. All plots show variation with embankment height, h. The four rows correspond to the values of a and s in the rows of Table 4.3. The columns show different values of reinforcement stiffness, J ($= 1.0$, 3.0 or 10 MN/m). In general higher J has been considered for larger clear spacing ($s - a$), and for the largest value of ($s - a$) only the highest J has been considered. The values of J represent a wide range of practical values, which are plausible in each case based on Eq. 4.4.

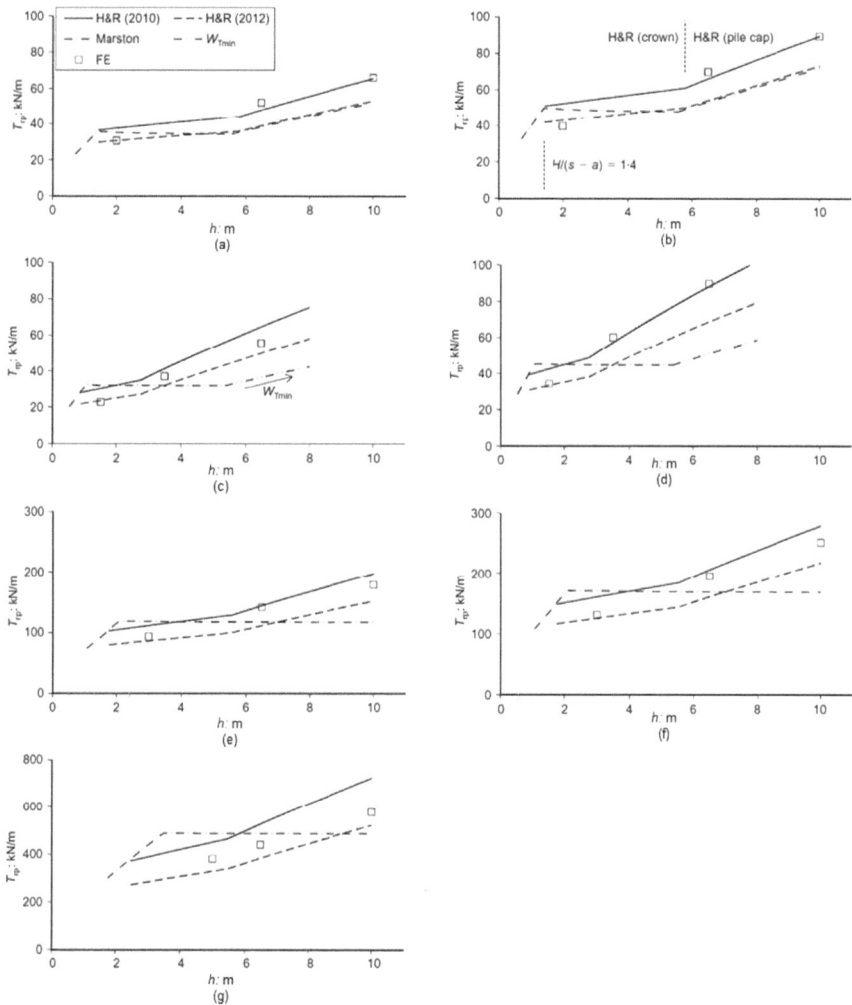

Fig. 4.11 Variation of reinforcement tension (T_{rp}) with embankment height (h): predictions and FE results **a** $a = 1.0$ m; $s = 2.0$ m; $J = 1.0$ MN/m; **b** $a = 1.0$ m; $s = 2.0$ m; $J = 3.0$ MN/m; **c** $a = 0.5$ m; $s = 1.25$ m; $J = 1.0$ MN/m; **d** $a = 0.5$ m; $s = 1.25$ m; $J = 3.0$ MN/m; **e** $a = 1.0$ m; $s = 2.5$ m; $J = 3.0$ MN/m; **f** $a = 1.0$ m; $s = 2.5$ m; $J = 10$ MN/m; **g** $a = 1.0$ m; $s = 3.5$ m; $J = 10$ MN/m

The predictions from Eq. 4.4 are shown as continuous variation with embankment height h. For direct comparison with the FE analyses, partial load factors (f) have been taken as 1, and the surcharge (w_s) is 0. For the Marston prediction of W_T the piles were assumed to be 'normal' (rather than 'unyielding'), noting that this gives a more conservative (higher) value (e.g. Ellis & Aslam, 2009). For the Hewlett and Randolph ('H&R') approach the passive earth pressure coefficient K_p was taken as 3.0, corresponding to $\phi' = 30°$ for the embankment fill in the FE analyses. 'H&R

(2010)' and 'H&R (2012)' refer to Eqs. 4.2a, b respectively, noting that σ_r follows directly from the original method. The W_{Tmin} limit ($= 0.15\gamma s H$ for $f = 1$ and $w_s = 0$) was also applied where relevant.

As illustrated in Fig. 4.11, changes in the response are observed for the Marston approach when $h/(s-a) = 1.4$ (Fig. 4.11b), and when the W_{Tmin} limit is applicable for higher embankments (Fig. 4.11c). As $(s-a)$ increases the W_{Tmin} limit does not have effect (Fig. 4.11e–g). For the Hewlett and Randolph approach there is a transition from criticality of failure at the crown of the arch to the pile cap as embankment height increases (Zhuang et al., 2012), and the W_{Tmin} limit does not take effect for any of the geometries considered. Predictions of T_{rp} from H&R (2012) are between 18 and 28% smaller than H&R (2010), where the difference increases with (s/a).

The increase in J between the two columns results in a reduction in implied reinforcement strain (Eq. 4.6). Taking a nominal allowable strain of 5%, T_{rp} is 50, 150 and 500 kN/m for $J = 1.0$, 3.0 or 10 MN/m respectively. The values predicted illustrate that the combinations of geometry and stiffness are broadly sensible. e.g. $T_{rp} > 50$ kN/m for large h in Fig. 4.11a, but $T_{rp} < 150$ kN/m for all h in Fig. 4.11b.

The FE results are shown by discrete data points at 3 values of h on each plot. In general agreement with the Hewlett and Randolph approach is good: The 2010 and 2012 approaches appear to give upper and lower bounds respectively. This is perhaps surprising since:

(a) The Hewlett and Randolph method overpredicts vertical stress in the arching embankment compared to FE analysis.
(b) Equation 4.2a is not consistent with vertical equilibrium.
(c) Equation 4.3 does not specifically account for non-uniform stress in the reinforcement, and is thus likely to underpredict T_{rp}.
(d) There is uncertainty regarding the relationship linking T_{rp} and W_T (e.g. Eqs. 4.2a, b).

In fact it would appear that the net effect is (perhaps fortuitously) approximately neutral compared to the FE analyses. It is also potentially concerning that the current revision of BS 8006 somewhat underpredicts the data. Concerns regarding underprediction from the Marston method as h increases appear to be confirmed (e.g. Ellis & Aslam, 2009; Van Eekelen et al., 2011), although the W_{Tmin} limit mitigates this effect to some extent. Agreement for the Marston method (and the W_{Tmin} limit) is certainly not as good as the Hewlett and Randolph method.

4.4.2.4 Comparison of Reinforcement Sag from FE Results and Prediction

BS 8006 () also suggests that the maximum sag in the reinforcement (δ_g) spanning between pile caps can be estimated as

$$\frac{\delta_g}{s-a} = \left(\frac{3\varepsilon}{8}\right)^{0.5} \tag{4.14}$$

This equation is based on a span $(s - a)$, rather than the diagonal which is $\sqrt{2}$ times longer. The maximum sag is (unsurprisingly) observed on the diagonal, and thus using $(s - a)\sqrt{2}$ in place of $(s - a)$, and again assuming that $\varepsilon = (T_{rp}/J)$

$$\frac{\delta_g}{s-a} = \sqrt{2}\left(\frac{3\varepsilon}{8}\right)^{0.5} = \left(\frac{3T_{rp}}{4J}\right)^{0.5} \tag{4.15}$$

BS 8006 actually suggests a factor of 2 (rather than $\sqrt{2}$) be applied to Eq. 4.14 when considering sag on the diagonal. This would give predictions of normalised sag $\sqrt{2}$ times higher than Eq. 4.15. Love and Milligan (2003) note that for separate orthogonal layers the maximum sag would result from addition of sag in the layers.

For reinforcement strain in the range 3–6%, Eq. 4.15 predicts normalised sag in the range 15–21%. Hence normalised maximum sag $\delta_g/(s - a)$ was predicted using Eq. 4.2a for W_T, Eq. 4.4 for T_{rp}, and Eq. 4.15. Figure 4.12 shows the result for each FE geometry (Table 4.3), compared with the corresponding FE results. Where two values of J were used for a given geometry, 'filled' (or 'solid') data points are used for the higher value, and thus these points have lower sag. The range of data spans (exceeds) the range of 15–21% mentioned above. Agreement is good, with error up to $\pm 20\%$ in the prediction from Eq. 4.15 compared to the FE results. As expected, using Eq. 4.2b instead of Eq. 4.2a led to a reduction (of 10–15%) in predicted sag, and thus agreement was not as good.

Fig. 4.12 Normalised maximum sag $y/(s - a)$: comparison of FE results and prediction using Eqs. (4.2a), (4.4) and (4.15)

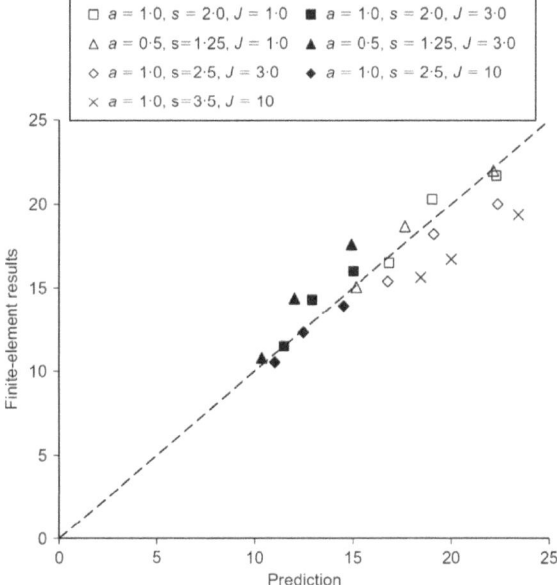

4.5 Summary

This chapter has focussed on the role of geogrid reinforcement in addition to the GRC for arching in the embankment previously considered in Chap. 3. The analyses were again conducted in plane strain and extended 3D condition for a square grid of pile caps. It was found that the geogrid was capable of reducing the ultimate stress on the subsoil to zero. However, this required significant sag of the geogrid. Comparison of the geogrid action with a simple formula (Eq. 2.30) for the sag gave reasonable agreement. Further development of the formulae indicated that the sag was very sensitive to the span of the geogrid between piles, but relatively insensitive to the stiffness of the geogrid (Eq. 4.12). This observation was supported by the results of the analyses.

For a case with three layers of geogrids, the upper two layers of grids carried relatively little tension compared to the bottom layer. This finding has been proposed by Jenner et al. (1998) in a field study. They stated that larger grid strains were recorded in the lower grid than the upper gird as anticipated in the design. In 3D case, it was found that the maximum tension in the geogrid occurred at the corner of the pile cap, contrasting with the plane strain result where tension was virtually constant across the span and pile cap. The tension on a diagonal between pile caps is about 2–3 times smaller than the maximum and is somewhat overpredicted by the plane strain formula using the diagonal span. However, a version of the formula modified to account for the concentration of load in the geogrid directly between the piles gave reasonable correspondence with the maximum tension at the corner of the pile cap.

BS 8006 (2012) allows prediction of reinforcement tension in a piled embankment using the 'Marston' and Hewlett and Randolph (1988) approaches as a basis for embankment load exerted on the reinforcement, and noting that the latter has been changed compared to BS 8006 (2010). This chapter also presents predictions by these methods for a wide range of piled embankment geometries, and comparison with results from three-dimensional Finite Element analyses. The analyses consider complete loss of subsoil support, and reinforcement. Concerns regarding use of the Marston approach for high embankments (e.g. Ellis & Aslam, 2009; Van Eekelen et al., 2011) are confirmed. However, the Hewlett and Randolph approach agrees well with the FE results for all geometries, with the BS 8006 2010 and 2012 approaches providing reasonable upper and lower bounds to the data respectively. This is perhaps surprising since it can be argued that individual component elements in the ultimate prediction appear somewhat deficient and inaccurate. It is also potentially concerning that the 2010 revision (upper bound to the data) is now superseded, and the current version is a lower bound. A suggested modification to the BS 8006 formula for prediction of reinforcement sag also gives good agreement with the FE analyses.

References

Almeida, M. S. S., Ehrlich, M., Spotti, A. P., & Marques, M. E. S. (2007). Embankment supported on piles with biaxial geogrids. *ICE Proceedings: Geotechnical Engineering, 160*(4), 185–192.

Briancon, L., & Simon, B. (2012). Performance of a pile-supported embankment over soft soil: A full-scale experiment. *ASCE Journal of Geotechnical and Geoenvironmental Engineering, 138*(4), 551–561.

BS8006-1. (2010). *Code of practice for strengthened/reinforced soils and other fills.* British Standards Institution.

BS8006-1. (2012). *Code of practice for strengthened/reinforced soils and other fills, incorporating Corrigendum 1.* British Standards Institution.

Ellis, E. A., & Aslam, R. (2009). Arching in piled embankments: Comparison of centrifuge tests and predictive methods—Part 1 of 2. In *Ground engineering* (34–38), June 2009.

Halvordson, K. A., Plaut, R. H., & Filz G. M. (2010). Analysis of geosynthetic reinforcement in pile-supported embankments. Part II: 3D cable-net model. *Geosynthetics International, 17*(2), 68–76.

Hewlett, W. J., & Randolph, M. F. (1988). Analysis of piled embankments. In *Ground engineering* (12–18), April 1988.

Jones, B. M., Plaut, R. H., & Filz, G. M. (2010). Analysis of geosynthetic reinforcement in pile-supported embankments. Part I: 3D plate model. *Geosynthetics International, 17*(2), 59–67.

Jenner, C. G., Austin, R. A., & Buckland, D. (1998). Embankment support over piles using geogrids. In *Proceedings 6th international conference on geosynthetics*, Atlanta, USA, 763–766.

Liu, H. L., Ng, C. W. W., & Fei, K. (2007). Performance of a geogrid-reinforced and pile-supported highway embankment over soft clay: Case study. *ASCE Journal of Geotechnical and Geoenvironmental Engineering, 133*(12), 1483–1493.

Love, J., & Milligan, G. (2003). Design methods for basally reinforced pile-supported embankments over soft ground. In *Ground engineering* (39–43), March 2003.

Plaut, R. H., & Filz, G. M. (2010). Analysis of geosynthetic reinforcement in pile-supported embankments. Part III: Axisymmetric model. *Geosynthetics International, 17*(2), 77–85.

Russell, D., & Pierpoint, N. (1997). An assessment of design methods for piled embankments. *Ground Engineering*, November 1997, pp. 39–44.

Van Eekelen, S. J. M., Bezuijen, A., & van Tol, A. F. (2011). Analysis and modification of the British Standard BS 8006 for the design of piled embankments. *Geotextiles and Geomembranes, 29*, 345–359.

Zhuang, Y., Ellis, E. A., & Yu, H. S. (2012). Technical note: Three-dimensional finite element analysis of arching in a piled embankment. *Geotechnique, 62*(12), 1127–1131.

Chapter 5
Reinforced Piled Embankment with Subsoil in Plane Strain and Three-Dimensional Conditions

5.1 Introduction

In Chap. 4, the authors considered predictions of reinforcement tension in a piled embankment based on British standard BS 8006 published in 2010 and the 2012 amended version, and compared the results with finite-element model predictions. In keeping with BS 8006, any contribution from the subsoil beneath the embankment was ignored.

Han and Gabr (2002) performed a numerical study on reinforced piled embankments, in which the underlying subsoil was also involved. However, an axisymmetric analysis was used. Stewart and Filz (2005) also considered the effect of subsoil using numerical analysis, concluding that this should be a factor in design, but without considering how this might be achieved. EBGEO (2011) and Van Eekelen et al. (2012) considered elastic response of the subsoil in analytical models. However, the models are complicated, and the potentially very important effect of the subsoil preconsolidation stress is not considered. The final stage firstly considered in this chapter for plane strain and three-dimensional analysis is the presence of the soft subsoil beneath a reinforced embankment. The idealised long-term behaviour of the subsoil is investigated here, in conjunction with the ground reaction curve and reinforcement. In the three-dimensional analyses, elastic and elasto-plastic behaviour of the subsoil will be both considered.

This chapter also considers the contribution of a lightly overconsolidated subsoil layer. It is assumed that there is no 'working platform' (granular) below the pile cap level. The water table in the subsoil is assumed to remain stable, since either of these factors would be likely to significantly reduce the ability of the subsoil to support the embankment. Simple modifications to the BS 8006 method for prediction of reinforcement tension are proposed to account for the subsoil contribution, which are then compared with the three-dimensional finite analyses (3D FE) results.

© The Author(s) 2025
Y. Zhuang, *Arching in Piled Embankments Including the Effects of Reinforcement and Subsoil*, https://doi.org/10.1007/978-981-96-1198-0_5

5.2   Calculation of Reinforcement Tension Including the Effect of Subsoil

In this section, the method to predict the reinforcement tension of geogrid including the effect of subsoil was discussed. Firstly, the solution for components of vertical stress at the base of the embankment was presented. Based on the determination approach of the distributed (vertical) load carried by the reinforcement between pile caps W_T derived in Chap. 4, then the reinforcement tension can be calculated.

Zhuang et al. (2012) reported three-dimensional Finite Element (3D FE) modelling of arching in an unreinforced piled embankment, dependent on key geometrical variables (Fig. 5.1).

It is assumed that the tensile reinforcement and subsoil act together to support the base of an arching embankment (Fig. 5.1a)

$$\sigma_a = \sigma_r + \sigma_s \qquad (5.1)$$

where σ_a is the vertical stress at the base of the arching embankment; σ_r is the vertical stress carried by the reinforcement; σ_s the vertical stress carried by the subsoil. The values will vary with plan location, but consideration of average values will satisfy equilibrium. Each of these three components will be considered in turn below.

Zhuang and Ellis (2014) concluded that the Hewlett and Randolph (1988) approach for determination of 'maximum arching' in the embankment (σ_a) shows the most promising in BS 8006, and it will be used here. Zhuang et al. (2012) noted that maximum arching is reached at relatively small subsoil settlement, hence variation of σ_a with settlement is not explicitly considered.

Proceeding to the reinforcement contribution from tensile reinforcement, Abusharar et al. (2009), Ellis and Aslam (2009), Ellis et al. (2010), and Zhuang et al. (2014) have suggested that σ_r can be expressed in the form

$$\sigma_r = A\left(\frac{\delta_g}{s-a}\right)^3 \qquad (5.2)$$

where δ_g is the maximum sag of the reinforcement between pile caps (normalised by the clear spacing between pile caps, $s - a$); A is a variable with units kN/m^2.

Finally, the 'elastic' subsoil response (limited by the preconsolidation stress) can be written as

$$\sigma_s = k_s \delta_s \leq \Delta\sigma_{vp} \qquad (5.3)$$

where δ_s is the settlement of the subsoil (which will vary with plan location, but will be taken as the maximum value, for compatibility with the maximum sag of the reinforcement, namely $\delta_g = \delta_s$); k_s is the 'subgrade reaction' at the surface of the subsoil (kN/m^2/m, e.g., van Eekelen et al., 2012); $\Delta\sigma_{vp}$ is the increment of stress on the subsoil to reach the preconsolidation stress.

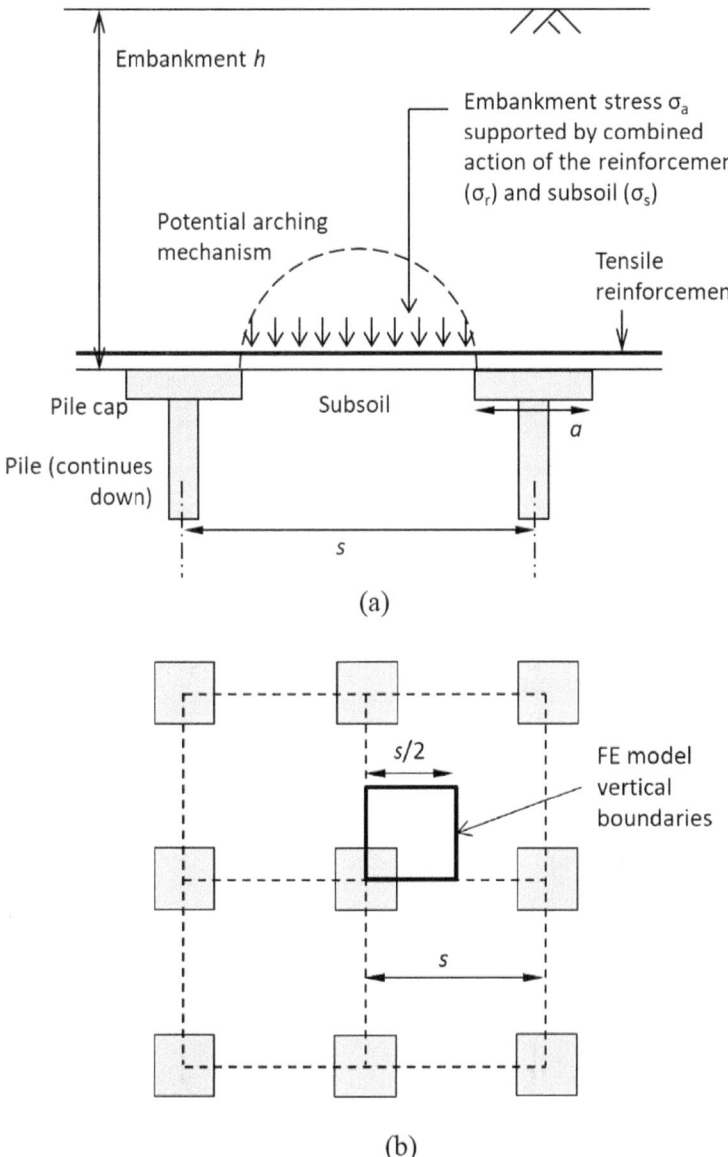

Fig. 5.1 Piled embankment illustrating behaviour and FE model boundaries; **a** schematic of behaviour and geometry (vertical section through pile caps), **b** FE model boundaries, one quarter of 'unit cell' (plan view)

To 'solve' Eq. 5.1, it is necessary to establish how σ_a (from the Hewlett and Randolph method) is distributed between σ_r and σ_s using compatibility of δ_g (Eqs. 5.2 and 5.3). Iteration is readily automated on a personal computer. If the limit $\sigma_s = \Delta\sigma_{vp}$ is reached in Eq. 5.3, then σ_r is independent of δ_g

$$\sigma_r = \sigma_a - \Delta\sigma_{vp} \tag{5.4}$$

Once σ_r has been determined, further analysis is required to determine the tension in the reinforcement. Chapter 4 discussed the distributed (vertical) load carried by the reinforcement between pile caps (W_T, kN/m) in BS 8006 (BSI, 2010, 2012). The corresponding equations will be written as

$$W_T = \alpha s \sigma_r \tag{5.5}$$

where $\alpha = 1$ (BSI, 2010); $\alpha = (s + a)/2\ s$ (BSI, 2012), noting that the 2010 version is not currently adopted, and is not consistent with vertical equilibrium, but that the 2012 version predicts lower load (since $\alpha < 1$). Then the reinforcement tension (T_{rp}, kN/m) can be obtained based on Eq. 4.4 as reported in Chap. 4.

5.3 Analyses Presented

The numerical models for the plane strain conditions and three-dimensional condition were introduced. The material parameters for subsoil, finite element mesh, boundary conditions, and summary cases of analyses all reported in detail.

5.3.1 For Plane Strain Condition

In this section, the numerical modelling of a geogrid-reinforced embankment with subsoil is again performed using Abaqus Version 6.6. Typical mesh geometry is shown in Fig. 5.2 (for $h = 3.5$ m, s = 2.5 m and subsoil thickness $h_s = 5.0$ m). There are 1566 eight nodded, reduced-integration, two-dimensional, quadratic solid elements (CPE8R) for the embankment, 16 three node quadratic truss elements (T2D3) for the reinforcement, and 1072 eight nodded, reduced-integration, two-dimensional, quadratic solid elements (CPE8R) for the subsoil.

The model now consists of reinforced embankment and subsoil. For the top reinforced embankment part, the vertical boundaries represent lines of symmetry at the centreline of a support (pile cap), and the midpoint between supports (see Sect. 2.1, Fig. 2.1) as before with corresponding restraint. The geogrid is again positioned 100 mm above the base of the embankment, with restraint on horizontal movement

Fig. 5.2 Typical finite element mesh ($h = 3.5$ m, $s = 2.5$ m, $h_s = 5$ m) and boundary conditions for reinforced embankment with subsoil

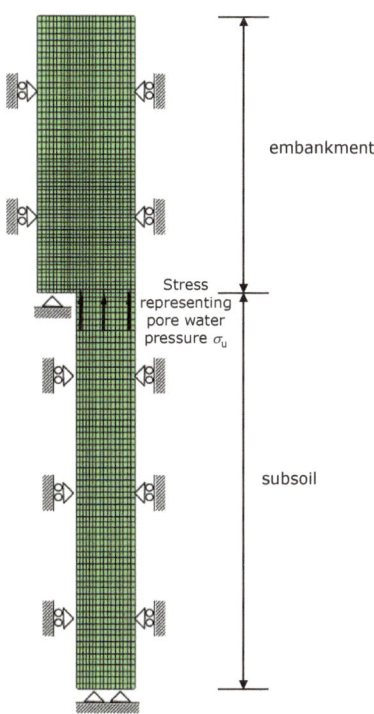

embankment

Stress representing pore water pressure σ_u

subsoil

at both sides. Again there are no boundary conditions imposed at the top embankment surface, and no surcharge is considered to act here.

The embankment is underlain by a half pile cap (width $a/2$) on the left, and subsoil (width $(s - a)/2$) at the right. For the subsoil, the vertical boundaries represent the edge of a pile, and the midpoint between piles. There is restraint on horizontal (but not vertical) movement at both boundaries. The bottom boundary represents the base of the soft subsoil, and there is rigid restraint on both vertical and horizontal movement. The subsoil is assumed to be underlain by a relatively stiff layer, and it is assumed that there is no settlement below the soft soil. It is effectively assumed that the pile has the same dimensions as the cap. This would not be the case in practice, but the analysis is somewhat idealised in this respect.

The pile cap is again assumed to provide rigid restraint to the embankment, and a vertical stress supporting the embankment (σ_u) is used to control the analysis. Initially σ_u is equal to the nominal overburden stress from the embankment (and the stress in the subsoil is zero). As σ_u is reduced the embankment material tends to arch and the reinforcement tends to sag as before. However, the subsoil also compresses with corresponding increase in vertical stress. Ultimately σ_u is reduced to zero, at which point the stress in the subsoil equals the stress at the base of the embankment (beneath the reinforcement).

This process is an idealisation, where reduction of σ_u is analogous to the effect of the excess pore water pressure in the subsoil, which initially carries the weight of the embankment, but is finally zero at the end of consolidation. The stress in the subsoil is initially zero, but increases as σ_u reduces, and thus actually represents the increase in vertical effective stress in the subsoil. However, this is sufficient to give a corresponding elastic settlement of the subsoil for one-dimensional conditions. σ_u is used to control the analysis in a similar way to σ_s in Chaps. 3 and 4. Now that the subsoil is actually modelled, σ_s is the stress at the top of the subsoil layer in the analysis.

The pile cap width (a) was fixed at 1 m and the centre-to-centre spacing (s) was 2.5 m. The embankment height (h) was 3.5 or 10 m. The thickness of subsoil (h_s) was 5 or 10 m. For the reinforcement installed at the base of the embankment, the distance between the top of the pile cap and the first layer of reinforcement (geogrid) is 0.1 m and the distance between multiple layers is again 0.3 m for three layers of geogrid (a Load Transfer Platform).

Throughout the analyses minimum and maximum element sizes of the embankment and subsoil were approximately 0.0006 and 0.0076 m^3/m. This corresponds to side lengths approximately in the range 30–150 mm. The length of truss elements representing the reinforcement was in the range 20–100 mm.

The embankment fill was again modelled as a linear elastic material, with a Mohr–Coulomb yield criterion. The parameters are the same as parameters used in Chap. 3 (see Table 3.1). The parameters of the geogrid are also the same as in Chap. 4 (see Table 4.1). The subsoil is considered as a linear elastic material with parameters shown in Table 5.1. The 'ambient' stress level does not affect the behaviour of this material and therefore the self-weight was not considered, and as above only the change in effective stress during consolidation was considered, giving a corresponding settlement.

All analyses in this chapter for plane strain condition are summarised in Table 5.2, where non-standard values are highlighted in bold. For $h = 3.5$ m, $s = 2.5$ m, $J = 6$ MN/m, $\delta_i = 0$ with a subsoil thickness $h_s = 5$ m, the effect of increasing Young's Modulus of subsoil E_s from 2.5 to 10 MN/m^2 was considered in cases (a–c). For $h = 3.5$ m, $s = 2.5$ m, $\delta_i = 0$ with subsoil thickness $h_s = 5$ m and $E_s = 5$ MN/m^2, the effect of increasing k from 6 to 12 MN/m, or using three geogrid layers with $J = 2$ MN/m was also considered in cases (d) and (e). The influence of increasing the geogrid interface friction angle δ_i to 20° was also considered (f). In cases (g) and (h) a thicker soft soil layer ($E_s = 2.5$ MN/m^2 and $h_s = 10$ m) was considered.

The in-situ stresses were specified for the reinforced embankment (again based on a unit weight of 17 kN/m^3 and a K_0 value of 0.5), with zero stress in the subsoil. Initially σ_u was specified as the nominal vertical stress at the base of the embankment

Table 5.1 Material parameters for subsoil

Young's modulus (MN/m^2)	Poisson's ratio
5 (or 2.5, or 10)	0.2

Table 5.2 Summary of analyses reported in plane strain condition

h (m)	s (m)	J (MN/m)	δ_i	Young's Modulus of subsoil E_s (MN/m^2)	Height of subsoil h_s (m)	Subplot
3.5	2.5	6	0	2.5	5	(a)
3.5	2.5	6	0	5.0	5	(b)
3.5	2.5	6	0	10.0	5	(c)
3.5	2.5	12	0	5.0	5	(d)
3.5	2.5	3×2	0	5.0	5	(e)
3.5	2.5	3×2	20	5.0	5	(f)
10.0	2.5	6	0	2.5	10	(g)
10.0	2.5	3×2	20	2.5	10	(h)

to give equilibrium with the in-situ stresses, but this value was then reduced to mimic consolidation of the subsoil as described above.

5.3.2 For Three-Dimensional Condition

A schematic of the FE model is respectively shown in Fig. 5.3. Figure 5.4 presented a typical geometry mesh for the FE model with embankment height (h) = 3.5 m, subsoil thickness (h_s) = 5.0 m, and pile spacing (s) = 2.5 m. There are 5328 eight nodded, full-integration, three-dimensional, linear brick solid elements (C3D8) for the embankment; 100 four noded, full-integration, three-dimensional membrane element (M3D4) for the geogrid reinforcement; and 6069 eight nodded, full-integration, three-dimensional, linear brick solid elements (C3D8) for the subsoil.

The vertical boundaries represent planes of symmetry passing through the centre of a pile cap and the midpoint between pile caps. The restraint on movement at these boundaries is as described in Chaps. 3 and 4, with the same restraint on movement normal to the faces in the subsoil as for the embankment. As in the equivalent plane strain analyses for convenience subsoil beneath the pile cap is not modelled. It is effectively assumed that there is a rigid inclusion with plan dimension the same as the pile cap through the full depth of the subsoil. Material beneath the subsoil is assumed to be rigid. As in plane strain analysis, the pile cap is assumed to provide rigid restraint to the embankment. The analysis was controlled by reduction of a stress σ_u applied at the embankment/subsoil interface. The initial value of σ_u was equal to the nominal vertical stress from the embankment, ultimately reducing to zero. Notionally this models the dissipation of excess pore pressure in the subsoil.

Table 5.3 summarises the analyses undertaken. The height of subsoil (h_s) was 3.5, 5 or 10 m. Meshes contained approximately 35–190 thousand elements, the minimum and maximum element size in the embankment and subsoil were approximately 0.0003 and 0.001 m^3 throughout the analyses. This corresponds to element dimensions of size 30–100 mm. The membrane element length and width were

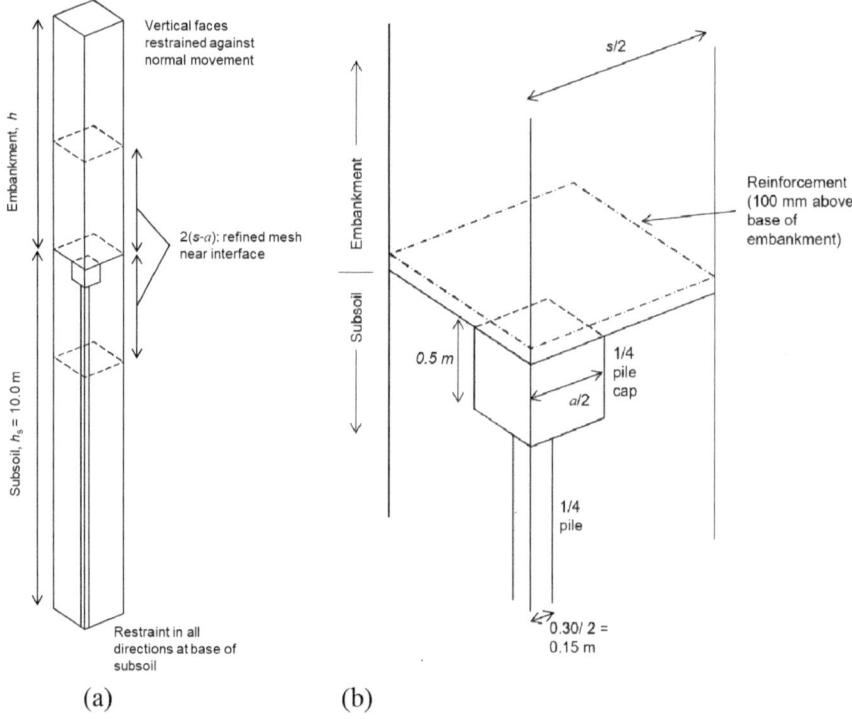

Fig. 5.3 Finite-element mesh geometry; **a** overview, **b** enlarged detail at embankment/subsoil interface

125 mm, and thickness was 1 mm. and the sensitivity to mesh size was checked similar to Zhuang and Ellis (2014).

Constitutive modelling of the embankment as a dry elastic-perfectly plastic material which was identical to previous chapters. The reinforcement (geogrid) is located at the bottom of embankment, where the thickness of the fill below the reinforcement and above the pile cap is again assumed to be 0.1 m in the model. Two further layers above the bottom geogrid are located layer by layer with 0.1 m gap between any two adjacent layers of geogrids. The geogrid is again modelled using three-dimensional membrane elements which can carry tension force but do not have any bending stiffness.

The thick soft subsoil was considered as a linear elastic material or using Modified Cam Clay model (MCC; Wood, 1990), its material parameter is shown in Tables 5.1 and 5.4. The hydrostatic groundwater pressure is calculated from the water table at the surface of the subsoil. It is noted that as the linear elastic subsoil was used, there was no requirement to model the self-weight of this material, and response of this layer was based on the increase in stress acting on it which occurred during the analysis. It was assumed that the clay had previously experienced a small vertical

Fig. 5.4 Typical finite
element mesh ($h = 3.5$ m,
$s = 2.5$ m, $h_s = 5$ m) for
reinforced embankment with
subsoil

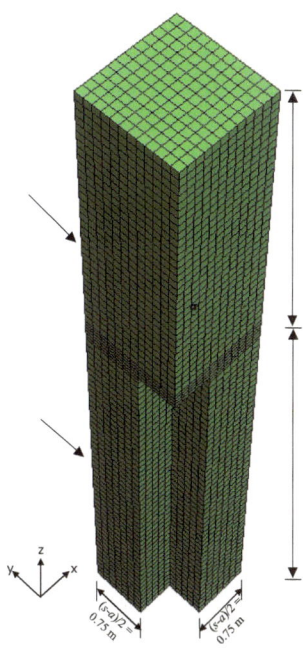

Table 5.3 Geometry and reinforcement stiffness

Pile cap, a: m	C–C cap spacing, s: m	Embankment height, h: m	Reinforcement stiffness, J: MN/m
1.0	2.0	2.0, 6.5, 10.0	1.0, 3.0
0.5	1.25	1.5, 3.5, 6.5	1.0, 3.0
1.0	2.5	3.0, 3.5,6.5, 10.0	2.0, 3.0, 10.0
1.0	3.5	5.0, 6.5, 10.0	10.0

'preconsolidation stress' ($\Delta\sigma_{vp}$) of 10 kN/m^2 in excess of the in-situ vertical effective stress.

The pile cap and pile were modelled as an elastic material (concrete) with Young's Modulus of 30 GN/m^2, and Poisson's Ratio of 0.20. They were assumed to be either 'smooth' or 'rough' in terms of the interface friction angle with the surrounding soil (δ). Smooth: $\delta = 0°$, Rough: $\delta = 10°$ for the subsoil (clay); $\delta = 15°$ for the (granular) embankment material (on the top face of the cap).

Table 5.4 Material
parameters of modified cam
clay model for subsoil

λ	κ	Γ	M
0.30	0.10	3.30	0.772

At the start of the analyses only the subsoil, pile cap and pile were present. The in-situ effective stress in the subsoil was specified based on the MCC parameters and pre-consolidation stress. The embankment was then constructed in layers, similar to Zhuang and Ellis (2012, 2014). During this process the soft clay subsoil was treated as 'undrained' (considered pragmatic and conservative). The clay subsoil was then allowed to consolidate via drainage at the top and bottom boundaries. Monitoring of pore water pressure confirmed that the anticipated excess pore pressure (γh) was generated during construction, and had completely dissipated by the end of the analysis (and hence the increase in total and effective stress in the layer were equal). Likewise, no sag or settlement occurred in the reinforcement during construction, but these values reached constant maxima by the end of consolidation.

5.4 Results

The results for the plane strain and three-dimensional conditions including behaviour of subsoil in different conditions (elastic subsoil and Modified cam clay subsoil), and reinforcement tension will be presented in the following section. Based on the numerical results, the contribution of subsoil and geogrid reinforcement for the equilibrium of the full system will be discussed. Furthermore, the accuracy of the proposed modified BS 8006 prediction will be evaluated compared to the finite-element results.

5.4.1 For Plane Strain Condition

Figure 5.5 shows the maximum displacement at the midpoint between the pile caps (δ_s) increasing with reduction in the stress at the bottom of reinforced embankment (σ_u).

Three lines are presented in Fig. 5.5:

- 'Embankment with geogrid'.
- 'Subsoil (analysis)'.
- 'Subsoil (comparison)'.

The 'embankment with geogrid' data comes from the previous analyses of a reinforced piled embankment, showing the initial data at relatively small settlement. The 'subsoil (analysis)' results are from the analyses presented in this chapter as the stress at the top of the subsoil as σ_u decreases, σ_s increases, and settlement δ_s increases.

The line 'subsoil (comparison)' is simply derived from one-dimensional compression theory. The one-dimensional modulus is given by

$$E_0 = E_y \frac{(1-\mu)}{(1+\mu)(1-2\mu)} \tag{5.6}$$

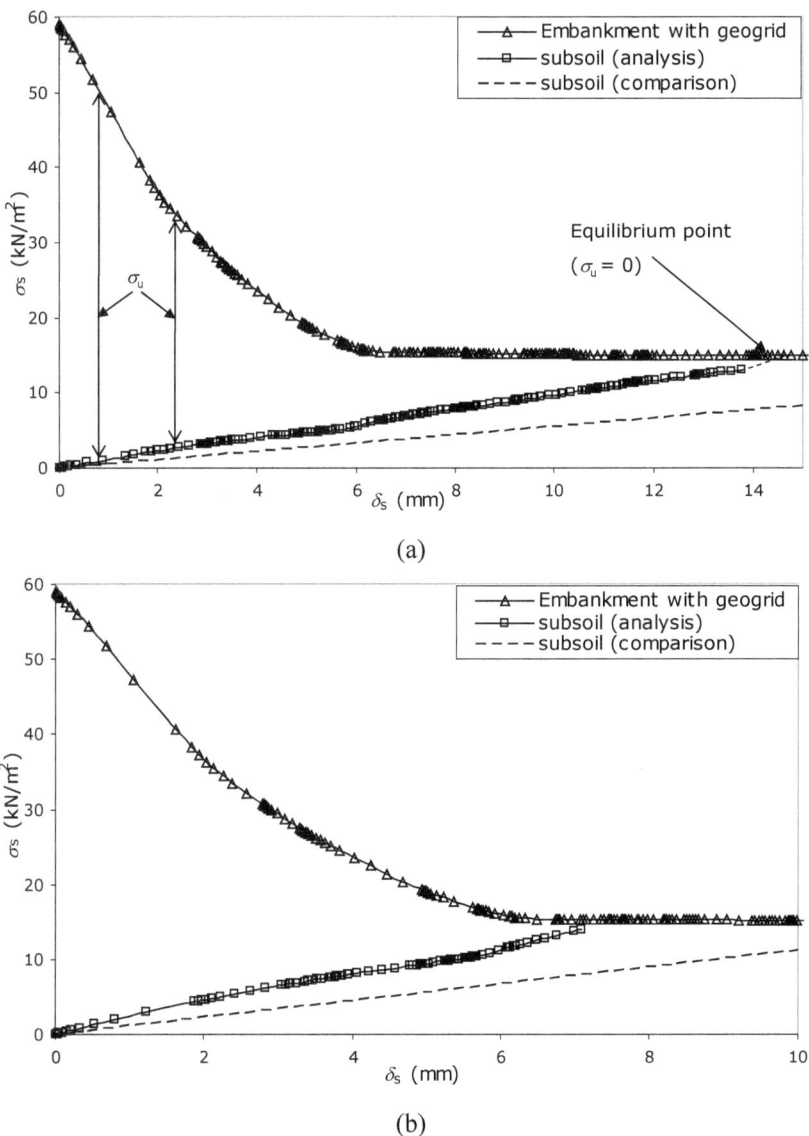

Fig. 5.5 Behaviour of subsoil in different conditions; **a** $h = 3.5$ m, $s = 2.5$ m, $J = 6$ MN/m, $\delta_i = 0$, $E_s = 2.5$ MN/m^2, $h_s = 5$ m, **b** $h = 3.5$ m, $s = 2.5$ m, $J = 6$ MN/m, $\delta_i = 0$, $E_s = 5$ MN/m^2, $h_s = 5$ m, **c** $h = 3.5$ m, $s = 2.5$ m, $J = 6$ MN/m, $\delta_i = 0$, $E_s = 10$ MN/m^2, $h_s = 5$ m, **d** $h = 3.5$ m, $s = 2.5$ m, $J = 12$ MN/m, $\delta_i = 0$, $E_s = 5$ MN/m^2, $h_s = 5$ m, **e** $h = 3.5$ m, $s = 2.5$ m, $J = 3 \times 2$ MN/m, with three layers of geogrid, $\delta_i = 0$, $E_s = 5$ MN/m^2, $h_s = 5$ m, **f** $h = 3.5$ m, $s = 2.5$ m, $J = 3 \times 2$ MN/m, with three layers of geogrid, $\delta_i = 20$, $E_s = 5$ MN/m^2, $h_s = 5$ m, **g** $h = 10$ m, $s = 2.5$ m, $J = 6$ MN/m, $\delta_i = 0$, $E_s = 2.5$ MN/m^2, $h_s = 10$ m, **h** $h = 10$ m, $s = 2.5$ m, $J = 3 \times 2$ MN/m, with three layers of geogrid, $\delta_i = 20$, $E_s = 2.5$ MN/m^2, $h_s = 10$ m

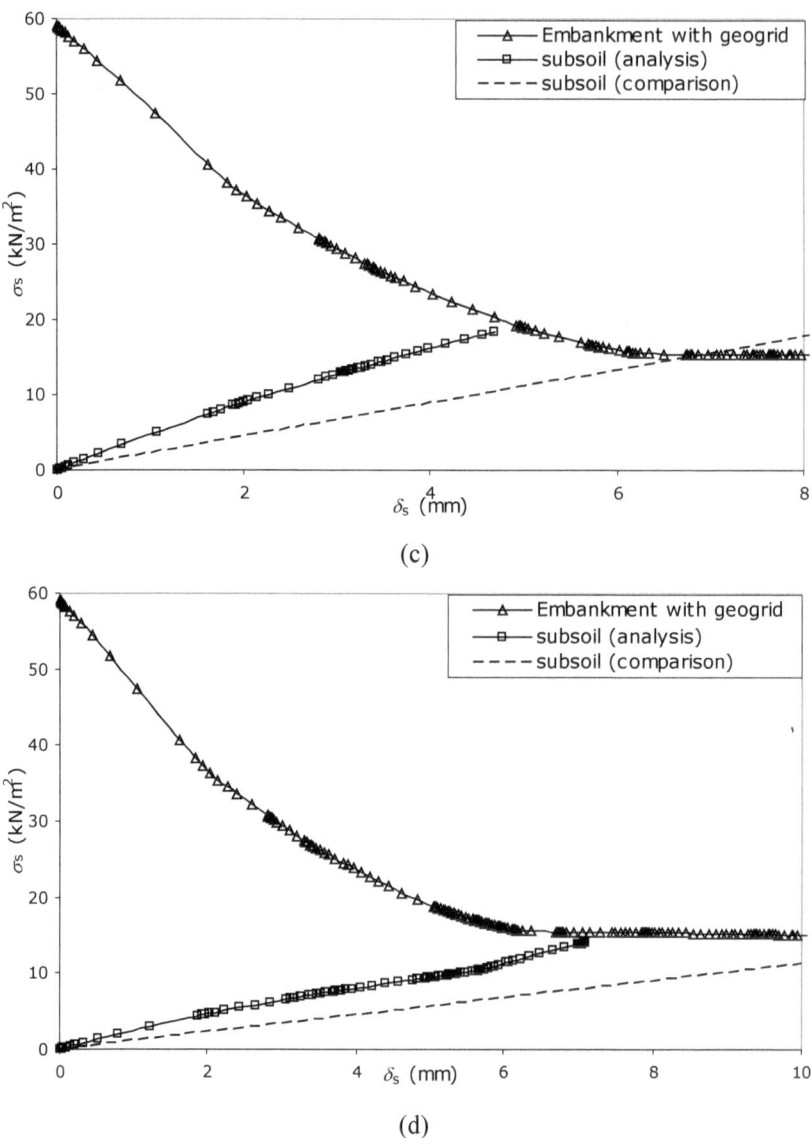

Fig. 5.5 (continued)

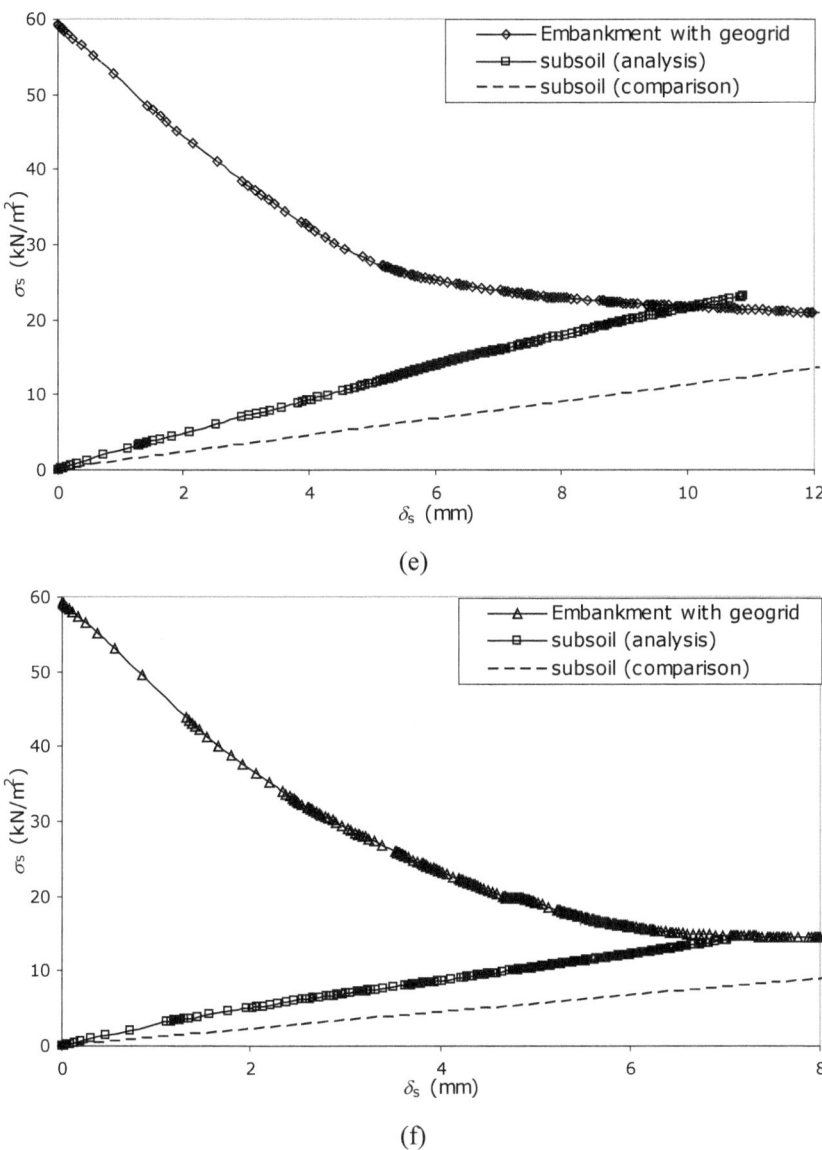

(e)

(f)

Fig. 5.5 (continued)

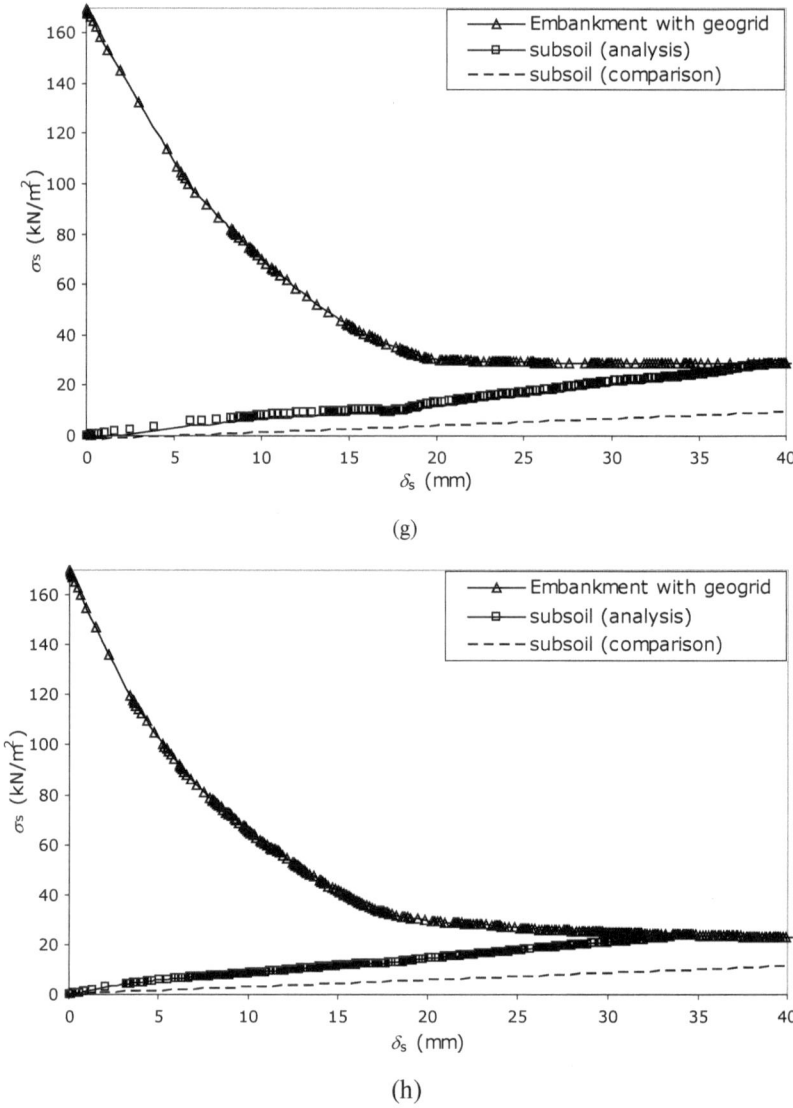

(g)

(h)

Fig. 5.5 (continued)

And then

$$\sigma_s = E_0' \frac{\delta_s}{h_s} \tag{5.7}$$

Fig. 5.6 Rotation of
principle stresses (subsoil)

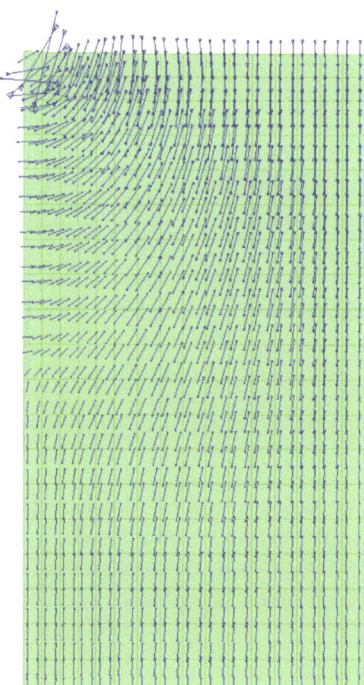

This gives a straight line through the origin on the chart, due to linear elastic response. As shown in Fig. 5.5, the 'subsoil (analysis)' line is steeper than the 'subsoil (comparison)' line; this effect appeared to be related to an additional effect of 'bearing failure' and associated rotation of principal stresses (see Fig. 5.6).

At the end of each analysis σ_u had reduced to zero. This logically corresponds to approximate intersection of the 'embankment with geogrid' and 'subsoil (analysis)' lines, since the stress at the base of the embankment is equal to the increase in stress in the subsoil. At this point the displacement is generally relatively small, and the geogrid carries very little load (compared with Chap. 4).

Comparing subplots (a), (b) and (c), the subsoil (analysis) line becomes steeper with increased Young's Modulus of the subsoil, and the settlement δ reduces from approximately 15 to 5 mm when σ_u and the lines intersect. Comparing subplots (b), (d) and (f) the displacement (δ) at the intersection point is approximately consistent with a value of 7 mm. This is because the geogrid has very little effect at this small displacement (see Fig. 4.9a).

Subplot (e) shows a slightly larger displacement than the subplot (f). This is because (as noted previously in Chap. 4) the mass strength at the base of the embankment has been reduced by the frictionless interfaces between embankment fill and the 3 geogrids.

For (g) and (h) the load on the geogrid is increased somewhat by increasing the embankment height to 10 m, reducing the subsoil stiffness to 2.5 MN/m^2 and increasing the subsoil thickness to 10 m. The settlement increases to approximately 35 mm, but the stress carried by the geogrid is still implied as small at this displacement (see Fig. 4.9a). Even in this situation the subsoil is stiffer than the geogrid and hence carries nearly all the remaining load at the point of maximum arching.

The stress carried by the subsoil is up to about 20 kN/m^2, which is higher than the value predicted by the one-dimensional settlement equation. This appears to be related to rotation of the principal stress (Fig. 5.6) related to a bearing capacity mechanism. This effect may be limited in a Tresca (rather than elastic) soil due to yielding. However, this is unlikely to have significant impact unless the strength is very low.

5.4.2 For Three-Dimensional Condition

The results for three dimensional analyses of a reinforced embankment including the subsoil and reinforcement are presented, including the variation of the subsoil settlement with the subsoil vertical stress increment. Like plane strain conditions, elastic and elasto-plastic behaviour of the subsoil are both considered.

5.4.2.1 Behaviour of the Elastic Subsoil

Figure 5.7 shows how the maximum displacement of the subsoil at the midpoint between the pile caps (δ_s) increases with a reduction in the stress at the bottom of reinforced embankment (σ_u). Like in Sect. 5.4.1 there are three lines presented in Fig. 5.7. They are 'Embankment with geogrid', 'subsoil (analysis)' and 'subsoil (comparison)'. The 'Embankment with geogrid' line comes from previous analyses of reinforced piled embankments (see Fig. 4.7). The data for 'subsoil (analysis)' shows results from the analyses presented in this Chapter. As for the plane strain conditions, the line 'subsoil (comparison)' comes from simple consideration of one-dimensional compression.

As in Sect. 5.4.1, the 'subsoil (analysis)' line is steeper than the 'subsoil (comparison)' line. For the plane strain analyses this effect appeared to be related to an additional effect of 'bearing failure' and associated rotation of principle stresses (see Fig. 5.6). As in the plane strain analyses, at the end of each analysis σ_u had reduced to zero. This logically corresponds to approximate intersection of the 'embankment with geogrid' and 'subsoil (analysis)' lines, since the stress at the base of the embankment is equal to the increase in stress in the subsoil (σ_s). At this point the displacement is generally relatively small, and the geogrid carries very little load, even when the subsoil layer is very soft and thick (i.e. very compressible).

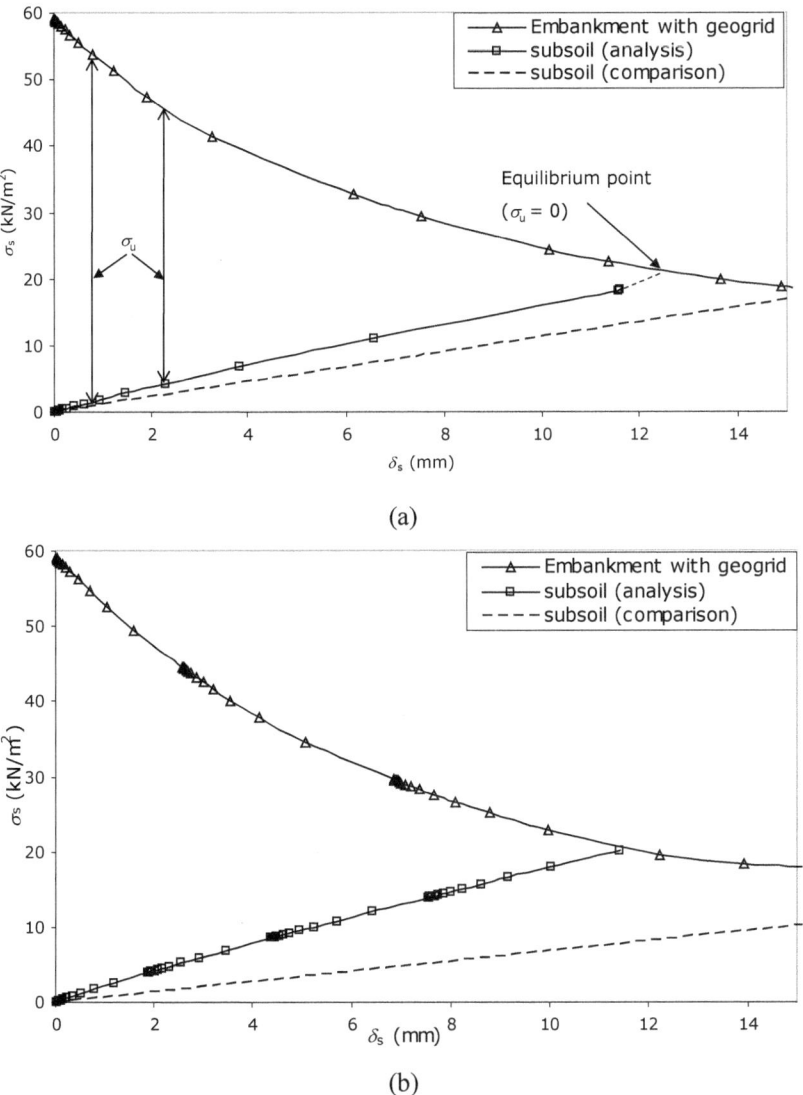

Fig. 5.7 Behaviour of subsoil in different conditions; **a** $h = 3.5$ m, $s = 2.5$ m, $J = 6$ MN/m, $\delta_i = 0$, $E_s = 5$ MN/m², $h_s = 5$ m, **b** $h = 3.5$ m, $s = 2.5$ m, $J = 3 \times 2$ MN/m, with three layers of geogrid, $\delta_i = 20$, $E_s = 5$ MN/m², $h_s = 5$ m, **c** $h = 10$ m, $s = 2.5$ m, $J = 6$ MN/m, $\delta_i = 0$, $E_s = 2.5$ MN/m², $h_s = 10$ m, **d** $h = 10$ m, $s = 2.5$ m, $J = 3 \times 2$ MN/m, with three layers of geogrid, $\delta_i = 20$, $E_s = 2.5$ MN/m², $h_s = 10$ m

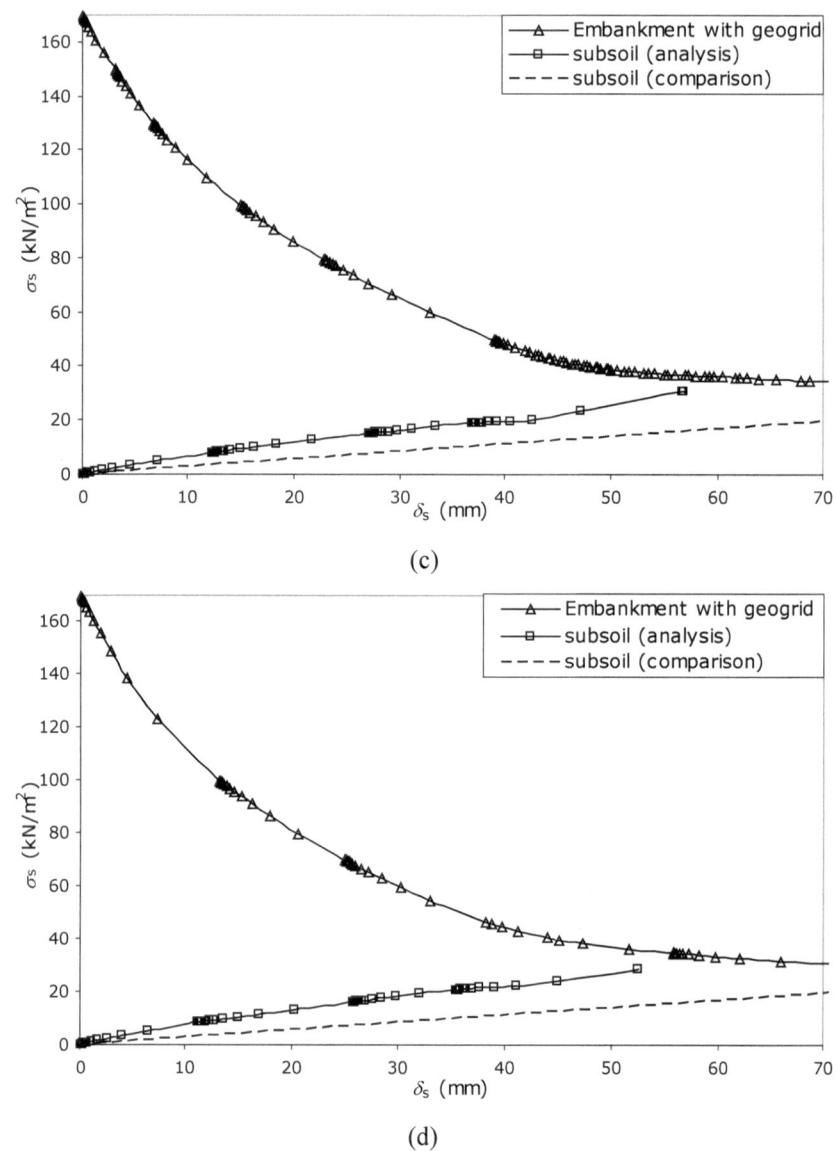

Fig. 5.7 (continued)

5.4.2.2 Behaviour of the Modified Cam Clay Subsoil

This section considers that the deformation behaviour of subsoil is simulated by the
Modified cam clay model (MCC model). Forty-two FE analyses are reported, from
the 21 combinations of geometry and reinforcement stiffness in Table 5.3, each for

smooth and rough piles. Figure 5.8 shows the subsoil vertical stress increment ($\Delta\sigma_v$) and settlement (δ_s) at the end of the analysis. As anticipated, increasing embankment height (h) and pile spacing (s) and caused $\Delta\sigma_v$ and δ_s to increase—data points for the largest s ($= 3.5$ m) are labelled in Fig. 5.8b.

'1-D prediction' lines are also shown, corresponding to purely vertical deformation, and uniform settlement with plan location. The 'elastic' one-dimensional subsoil stiffness ($E_0\prime$) was inferred from the MCC parameters, increasing approximately linearly from 0.20 MN/m² at the top of the subsoil layer to 2.55 MN/m² at

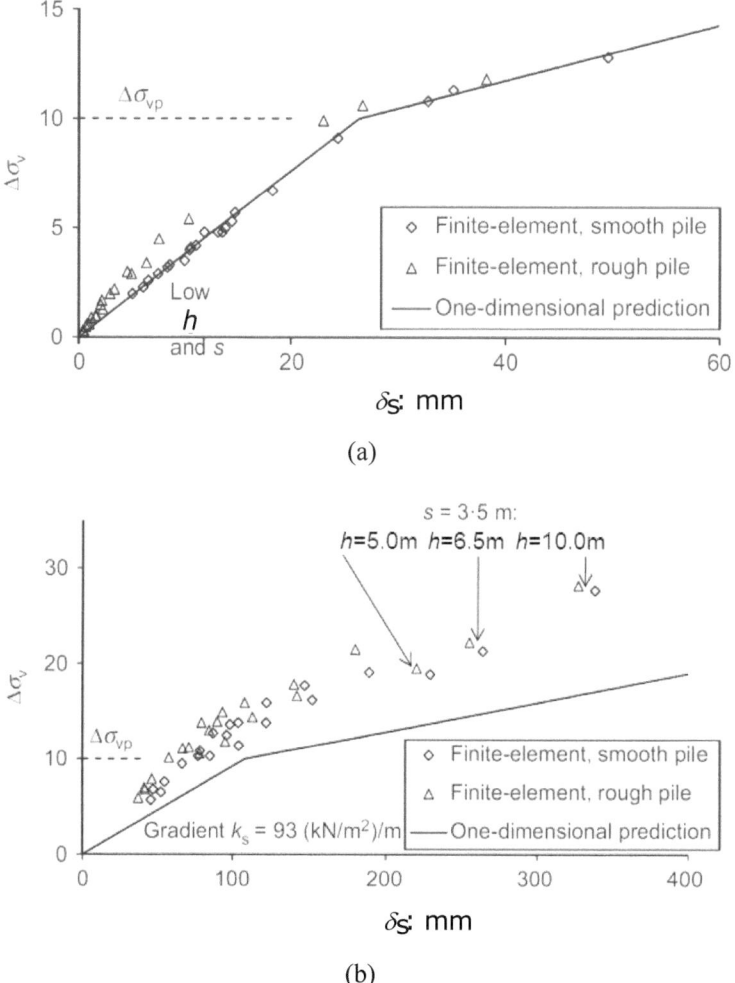

Fig. 5.8 Subsoil settlement (δ_s) and stress increment ($\Delta\sigma_v$); **a** $z = 5$ m (subsoil mid-depth), **b** $z = 0$ m (subsoil surface, maximum values at centre of diagonal span)

the base. For $\Delta\sigma_v > \Delta\sigma_{vp}$ a nominal reduction in stiffness is shown, corresponding to the ratio $\lambda/\kappa = 3$.

Figure 5.8a shows the response at mid depth of the subsoil (5.0 m), where settlement was uniform with plan location (except very locally near a rough pile). The smooth pile results show good agreement with the '1-D prediction', whilst the rough pile results are slightly stiffer (due to the beneficial effect of negative skin friction in this respect).

In contrast, settlement was highly non-uniform near the surface of the subsoil layer. $\Delta\sigma_v$ and δ_s were both maximum at the centre of a diagonal span between pile caps, and tended to zero at the pile cap. Figure 5.8b shows the response at the surface of the subsoil, now plotting the maxima of $\Delta\sigma_v$ and δ_s. The '1-D prediction' line now corresponds to Eq. 5.3, but noting that these equations were based on the concept average stress. The maxima of $\Delta\sigma_v$ and δ_s are somewhat stiffer than the 1-D prediction, and this was attributed to the localised shear strain resembling a bearing capacity mechanism observed near the surface of the subsoil. Again the rough pile gives stiffer response, but the effect is now relatively modest.

5.4.2.3 Reinforcement Tension

In Chap. 4, the reinforcement tension was analysed for the subsoil modelled via boundary conditions, which was also investigated in this chapter for the subsoil explicitly modelled. The general distribution of tension in the reinforcement was similar to Zhuang and Ellis (2014). Figure 5.9 shows the increase in maximum reinforcement tension (T_{rp}) with embankment height (h).

A total of four prediction lines are shown, and denoted as follows (e.g., 'H&R (2010) NSS'): 'H&R' is Hewlett and Randolph (1988) predictions of embankment arching, used throughout. '2010' and '2012' refer to revisions of BS 8006 (BSI, 2010, 2012) in Eq. 5.5. 'NSS' denotes 'No Subsoil Support', whereas 'SS' denotes 'Subsoil Support' (Eq. 5.3). SS implies use of Eq. 5.3, with $k_s = 93$ kN/m^2/m (based on the inferred subsoil E0' profile), or $\Delta\sigma_{vp} = 10$ kN/m^2 in Eq. 5.4. In Fig. 5.8b some FE results indicate $\Delta\sigma_v > \Delta\sigma_{vp}$, but a pragmatic design approach would ignore this.

In fact, the $\Delta\sigma_{vp} = 10$ kN/m^2 limit was reached in the majority of prediction cases. However, it was not in Fig. 5.9b, d, where the reinforcement response was relatively 'stiff' for low s and great J, and hence load on the subsoil was reduced.

Three sets of FE data points are shown and denoted as follows (e.g., 'FE SS sm'): 'NSS' ('No Subsoil Support') again refers to the results from Chap. 4, whereas 'SS' ('Subsoil Support') refers to the new analyses reported in this chapter; the SS analyses are either 'sm' or 'ro', referring to a 'smooth' or 'rough' pile respectively.

As anticipated, the effect of subsoil support is to reduce both the predictions of reinforcement tension and the FE results (where the rough pile slightly enhances this effect). The FE data is between 4 and 50% lower than the BS 8006 (2010) prediction including the proposed modification for subsoil support. Meanwhile the FE data is between 41% lower and 15% higher than the BS 8006 (2012) prediction including

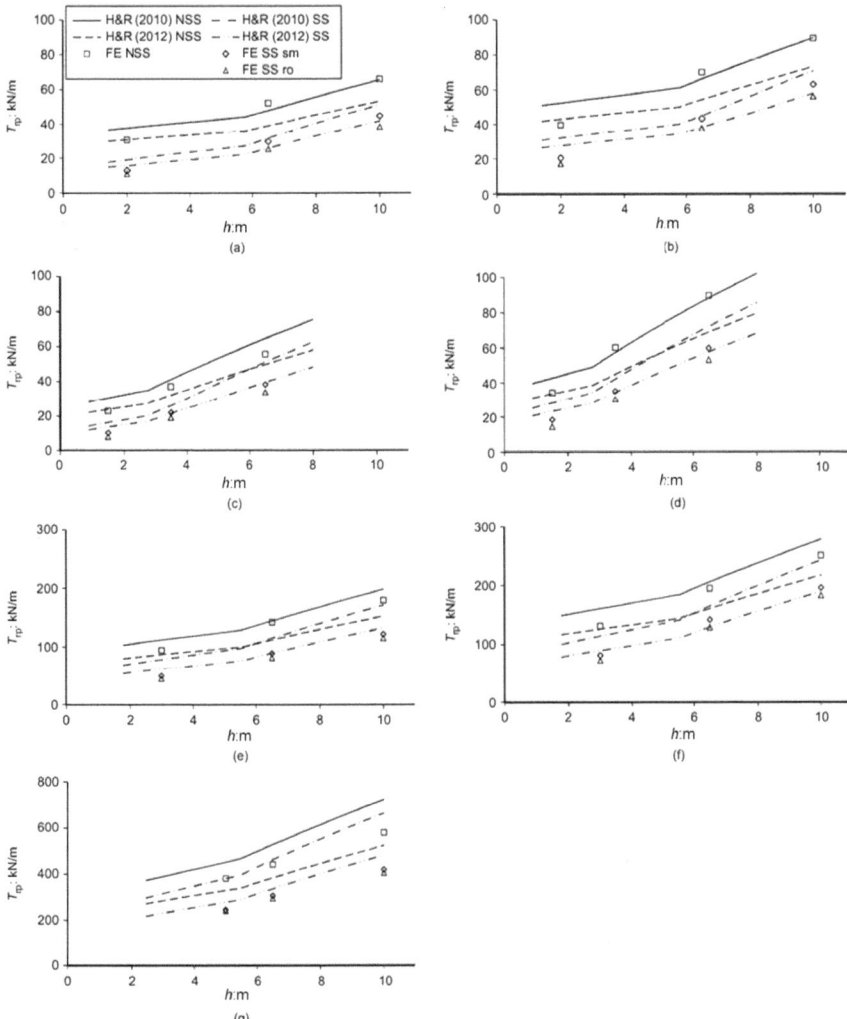

Fig. 5.9 Variation of reinforcement tension (T_{rp}) with embankment height (h): BS8006 predictions (including proposed modification for subsoil) and FE results: **a** $a = 1.0$ m, $s = 2.0$ m, $J = 1.0$ MN/m; **b** $a = 1.0$ m, $s = 2.0$ m, $J = 3.0$ MN/m; **c** $a = 0.5$ m, $s = 1.25$ m, $J = 1.0$ MN/m; **d** $a = 0.5$ m, $s = 1.25$ m, $J = 3.0$ MN/m; **e** $a = 1.0$ m, $s = 2.5$ m, $J = 3.0$ MN/m; **f** $a = 1.0$ m, $s = 2.5$ m, $J = 10$ MN/m; **g** $a = 1.0$ m, $s = 3.5$ m, $J = 10$ MN/m

the proposed modification for subsoil support. Where the FE data is significantly lower than the prediction (expressed as a percentage), this is for low tension (and low h).

5.5 Summary

This chapter extended the previous work by considering the potentially beneficial contribution of a clay subsoil layer, both in the finite-element predictions and as a simple modification to the BS 8006 predictive method. As anticipated, the subsoil support reduces reinforcement tension.

For the results in plane strain and three-dimensional conditions both show that the effect of the subsoil in the analyses was somewhat underestimated by a one-dimensional settlement prediction, due to the additional effect of principle stress rotation. The analyses indicate that the contribution to vertical equilibrium from the subsoil is considerably more significant than the geogrid. Hence equilibrium of the full system including arching in the embankment, geogrid reinforcement and subsoil is achieved at much lower settlement than the embankment and geogrid alone as shown in Chap. 4.

When compared with the finite-element results the proposed modified BS 8006 prediction is quite accurate. As noted in Zhuang and Ellis (2014) there are arguments that the component Equations under- and over-predict various aspects of the behaviour. Nevertheless, the ultimate prediction of T_{rp} (including modification for subsoil support) is quite good compared to the FE data. The BSI, 2012 modified prediction is the 'best' (but sometimes slightly unconservative), whereas the BSI, 2010 modified prediction is conservative in all cases considered in this chapter.

References

Abusharar, S. W., Zheng, J. J., Chen, B. G., & Yin, J. H. (2009). A simplified method for analysis of a piled embankment reinforced with geosynthetics. *Geotextiles and Geomembranes, 27*, 39–52.

BSI. (2010). *BS 8006-1: Code of practice for strengthened/reinforced soils and other fills*. British Standards Institution.

BSI. (2012). *BS 8006-1: Code of practice for strengthened/reinforced soils and other fills, incorporating Corrigendum 1*. British Standards Institution.

EBGEO. (2011). Recommendations for design and analysis of earth structures using geosynthetic reinforcements.

Ellis, E. A., & Aslam, R. (2009). Arching in piled embankments: Comparison of centrifuge tests and predictive methods—Part 2 of 2. *Ground Engineering, 42*.

Ellis, E. A., Zhuang, Y., Aslam, R., & Yu, H. S. (2010). Discussion of 'a simplified method for analysis of a piled embankment reinforced with geosynthetics' by Abusharar, Zheng, Chen and Yin, geotextiles and geomembranes 27, 39–52. *Geotextiles and Geomembranes, 28*, 133–134.

Han, J., & Gabr, M. A. (2002). Numerical analysis of geosynthetic reinforced and pile-supported earth platforms over soft soil. *Journal of Geotechnical and Geoenvironmental Engineering, 128*(1), 44–53.

Hewlett, W. J., & Randolph, M. F. (1988). Analysis of piled embankments. *Ground Engineering, 21*.

Love, J., & Milligan, G. (2003). Design methods for basally reinforced pile-supported embankments over soft ground. *Ground Engineering, 39*.

Stewart, M. E., & Filz, G. M. (2005). Influence of clay compressibility on geosynthetic loads in bridging layers for column-supported embankment. In *Contemporary issues in foundation engineering* (pp.1–14).

Van Eekelen, S. J. M., Bezuijen, A., Lodder, H. J., & van Tol, A. F. (2012). Model experiments on piled embankments, part II. *Geotextiles and Geomembranes, 32*, 82–94.

Wood, D. M. (1990). *Soil behaviour and critical state soil mechanics*. Cambridge University Press.

Zhuang, Y., & Ellis, E. A. (2014). Finite-element analysis of a piled embankment with reinforcement compared with BS 8006 predictions. *Geotechnique, 64*(11), 910–917.

Zhuang, Y., Ellis, E. A., & Yu, H. S. (2012). Technical note: Three-dimensional finite-element analysis of arching in a piled embankment. *Geotechnique, 62*(12), 1127–1131.

Zhuang, Y., Kang, Y. W., & Liu, H. L. (2014). A simplified model to analyze the reinforced piled embankments. *Geotextiles and Geomembranes, 42*, 154–165.

Chapter 6
Discussion of Results

6.1 Introduction

The aim of this chapter is to summarise the main findings of the finite element analyses, and then present some key comparisons. Finally, four case studies will be introduced.

6.2 Summary of Results

6.2.1 Piled Embankment

In this monograph, a series of plane strain and three-dimensional finite element analyses have been undertaken to investigate the behaviour of arching in piled embankments. The subsoil was not modelled, but the 'subsoil stress' at the base of the embankment was used to control the analysis. A parametric study mainly considered variation of the embankment height (h) and centre-to-centre pile spacing (s) with fixed pile cap dimension (a). Some investigation of the embankment soil frictional strength and dilation were also undertaken in plane strain.

The results showed that ratio of the embankment height to the centre-to-centre pile spacing (h/s) is a key parameter. For embankments with a value of h/s up to about 0.5, there is no evidence of arching based on the stress on the subsoil and there is significant differential settlement. As h/s increases from 0.5 to 1.5, the stress acting on the subsoil reduces compared to the nominal overburden stress from the embankment fill, which implies increasing effect of arching. The differential displacement at the surface of the embankment also reduces. For h/s larger than 1.5, the stress acting on the subsoil is considerably reduced compared to the nominal overburden stress, and there is no differential displacement at the surface of the embankment, i.e. there is 'full arching' developed.

© The Author(s) 2025
Y. Zhuang, *Arching in Piled Embankments Including the Effects of Reinforcement and Subsoil*, https://doi.org/10.1007/978-981-96-1198-0_6

6.2.2 'Reinforced' Piled Embankment

The effect of uniform biaxial reinforcement (geogrid or geotextile) in piled embankments was also studied in a series of plane strain and three-dimensional finite element analyses. During the study, the effects of the geogrid stiffness (J), the number of geogrid layers, and the interface friction angle were considered. Separate layers of uniaxial grid or 'primary' and 'secondary' reinforcement were not considered.

As expected, the reinforcement was capable of reducing the ultimate stress on the subsoil beneath the embankment (which again controlled the analysis) to zero. The sag of reinforcement could be very large, and was very sensitive to the span of the reinforcement between piles, but relatively insensitive to its stiffness. For the case with three layers of reinforcement distributed through the bottom metre of the embankment ('a load transfer platform'), the upper two layers carried relatively little tension compared to the bottom layer since they exhibited less sag. This trend of behaviour was also noted by Jenner et al. (1998) in a field case.

In the two-dimensional analyses the tension in the reinforcement was approximately constant across the span. However, in three-dimensional analyses, the results showed that the maximum tension in the geogrid occurred at the corner of the pile cap. Maximum reinforcement tension at the edge of the pile cap has also been noted by Russell and Pierpoint (1997).

6.2.3 'Reinforced' Piled Embankment with Subsoil

The effect of the soft subsoil in the 'reinforced' piled embankment was also presented in a series of plane strain and three-dimensional analyses (rather than just considering the stress acting at the surface of the subsoil). Variation of Young's Modulus and the thickness of the subsoil were considered in the study.

The results showed that the contribution to vertical equilibrium of the subsoil is generally more significant than the reinforcement for the cases considered. It was also found that the subsoil response was somewhat underestimated by consideration of the 1D stiffness since there was also a component of 'bearing' resistance in the soil immediately beneath the embankment.

6.3 Comparison of General Trends of Behaviour as $h/(s - a)$ Varies

The ratio of the embankment height (h) to the clear spacing between adjacent pile caps ($s - a$) has been considered as an important parameter for arching behaviour in design.

Aslam (2008) and Ellis and Aslam (2009a, 2009b) investigated the performance of unreinforced piled embankments supported by a square (3D) grid of piles in a series centrifuge tests. Their findings showed that:

- $h/(s-a) < 0.5$: there is no evidence of arching.
- $0.5 < h/(s-a) < 2.0$: there is increasing evidence of arching as h increases.
- $2.0 < h/(s-a)$: there is 'full' arching.

Potts and Zdravkovic (2008) performed finite element analyses to study the behaviour of geosynthetic reinforced fills overlying voids. They proposed that the development of arching in the fill depends on the ratio h/D (where h is the depth of overlying fill, and D is the width of the void), as well as the geometry of the void and the properties of the soil and geosynthetic. In all cases stable arching behaviour was found to occur when $h/D > 3.0$ for an infinitely long void (plane strain condition). Comparing the gap between piles in a piled embankment with a void h/D, is broadly equivalent to $h/(s-a)$.

In this work the results have considered the ratio h/s. However, this can be related to $h/(s-a)$:

$$\frac{h}{s-a} = \frac{h}{s}\frac{s}{s-a} \tag{6.1}$$

In this work the ratio (s/a) is in the range 2.0–3.5. $s/(s-a)$ is then in the range 2.0–1.4. Thus the critical value of $h/s = 1.5$ for full arching reported in this work corresponds to $h/(s-a) \approx 2.0 \sim 3.0$. This is consistent with the values reported by the other authors for physical and numerical modelling.

6.4   Comparison of the Value of $\sigma_s/\gamma(s-a)$ at the Point of Maximum Arching for Medium Height Embankments

Ellis and Aslam (2009a) presented plots of $\sigma_s/\gamma(s-a)$ against $h/(s-a)$. The results show that the value of $\sigma_s/\gamma(s-a)$ at the point of maximum arching is approximately 0.5. However, it was not possible to reliably determine σ_s for high embankments where the efficacy tended to 1.0.

Potts and Zdravkovic (2008) showed a plot of the magnitude of the vertical stress at the level of the reinforcement (which is equivalent to the subsoil stress, σ_s, in an unreinforced embankment as referred to in this work) against void diameter (D) for circular voids up to 4.0 m wide, Fig. 6.1. The results show the gradient of the 'ICFEP' line (from the finite element analyses) σ_s/D is approximately 5 kPa/m. Assuming $\gamma = 16$ kN/m^3, then $\sigma_s/\gamma D = 0.3$. The value D is analogous to $(s-a)$, and hence this value is approximately consistent with that proposed by Ellis and Aslam (2009a). It is perhaps not that surprising that the value (0.3) is somewhat lower in the axisymmetric case compared to arching over a square grid of piles (0.5).

Fig. 6.1 Comparison of vertical stresses at the level of the reinforcement (Potts & Zdravkovic, 2008)

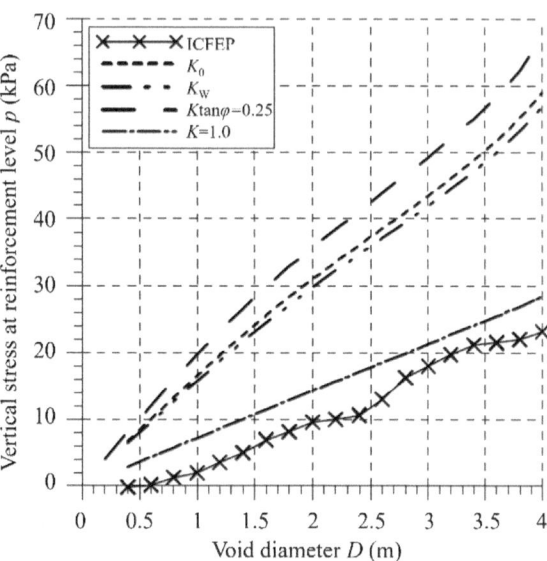

In this research, as shown in Figs. 3.6b and 3.11, the value of $\sigma_s/\gamma s$ for maximum arching is approximately 0.25–0.4 except for high s/a in the 3D situation, where punching of the pile caps into the base of the embankment gives higher values of $\sigma_s/\gamma s$. The results from this study can again be converted from normalisation by s to $(s - a)$ by multiplication by $s/(s - a) \approx 1.4$–2.0. This gives values of $\sigma_s/\gamma(s - a)$ in the range 0.4–0.8, which is consistent with the other research.

6.5 Equation for Equilibrium Including Arching, Reinforcement and Subsoil

Aslam (2008) and Ellis and Aslam (2009b) propose an interaction diagram where the combined action of arching in the embankment, reinforcement membrane action and the subsoil give equilibrium at a compatible settlement (δ)—see Fig. 2.18 (Sect. 2.4.3). δ corresponds to the maximum sag of the geogrid, but is considered as a uniform settlement at the surface of the subsoil.

Taking these individual components and writing the equation of equilibrium:

$$E_0 \frac{\delta}{h_s} + 21 \frac{J}{l} \left(\frac{\delta}{l}\right)^3 = \sigma_a + \sigma_w \tag{6.2}$$

where

$l = (s - a)$, clear spacing between pile caps (m)
$E_0 =$ the one-dimensional stiffness of the subsoil (kN/m^2)
$h_s =$ the thickness of the subsoil (m).

J = the stiffness of the geogrid (kN/m).

σ_a = the stress at the base of the embankment due to the action of arching alone (i.e. from the Ground Reaction Curve).

$\sigma_w = \gamma_w h_w$ the stress acting on the subsoil due to the working platform (any imported material below the pile cap level, which is hence not affected by arching).

The first term represents the subsoil response; the second is the geogrid membrane action, whilst σ_a and σ_w are the load that must be carried. In fact the geogrid cannot carry weight from the working platform (which is below it), and hence this term cannot exceed σ_a. Likewise the subsoil term cannot be less than σ_w since the working platform is only supported by the subsoil. At this point a gap would open between the reinforcement and working platform beneath.

It has been shown by Ellis et al. (2009) that this Equation is consistent with a similar but more complex equation proposed by Abusharar et al. (2009). The equation contains four extra terms, but these are shown to be relatively insignificant. The equation is only presented for a plane strain situation by Abusharar et al. (2009).

The span l to be used in the geogrid term has been unclear, particularly for a 3D pile cap layout. However, Chap. 4 indicates the following values for uniform biaxial reinforcement:

- 2D: $l = (s - a/2)$
- 3D: $l = \sqrt{2}\,(s - a)$

Hence for a 3D arrangement, the equation of equilibrium is as follows:

$$E_0\frac{\delta}{h_s} + 5\frac{J}{s-a}\left(\frac{\delta}{s-a}\right)^3 = \sigma_a + \sigma_w \tag{6.3}$$

If $\sigma_a = A\gamma(s - a)$ and the embankment and working platform have the same unit weight then the equation can be written in a non-dimensional form as:

$$\frac{E_0}{\gamma h_s}\frac{\delta}{s-a} + 5\frac{J}{\gamma(s-a)^2}\left(\frac{\delta}{s-a}\right)^3 = A + \frac{h_w}{s-a} \tag{6.4}$$

6.6 Case Studies

6.6.1 Second Severn Crossing

The Second Severn Crossing provides a second motorway link between South Wales and England across the River Severn estuary. The new toll plaza for the Second Severn Crossing was constructed on low lying land adjacent to the estuary. To alleviate the risk of flooding, ground levels were generally raised by between 2.5 and 3.5 m increasing locally to 6 m maximum height. The case study of this project has been provided by Maddison et al. (1996).

Table 6.1 Summary of subsoil properties for the second Severn crossing

	Thickness, t (m)	Stiffness, E_0(kN/m^2)
Desiccated clay	1–2	5000
Estuarine clay	2–3	1800
Peat	2–4	500

The ground investigation indicated soft subsoil to depths up to 8 m overlying sands and gravels and Trias sandstone. Table 6.1 summarises the soft subsoil properties.

Ground improvement comprising vibro concrete columns (VCCs) and a 'load transfer platform' (LTP) incorporating relatively low strength geogrids was used to support the embankment. In the design, the VCCs were installed on a triangular grid of 2.7 m maximum spacing founding in the sand and gravel deposits. The load transfer platform at the base of the embankment comprised granular fill incorporating two layers of Tensar SS2 geogrid, in order to promote arching in the granular fill and transfer the embankment loads onto the columns. The properties of the geogrid are shown in Table 6.2, which are derived from the short-term quality control strength at approximately 10% strain as reported by Maddison et al. The long-term stiffness would be lower than the value derived from the short-term quality control tests (for both cases). A cross section of the design is shown in Fig. 6.2. More information can be found in Maddison et al. (1996).

Table 6.2 Summary of SS2 geogrid properties for the second Severn crossing

Property	Transverse direction	Longitudinal direction
Short-term stiffness (kN/m)	300	150

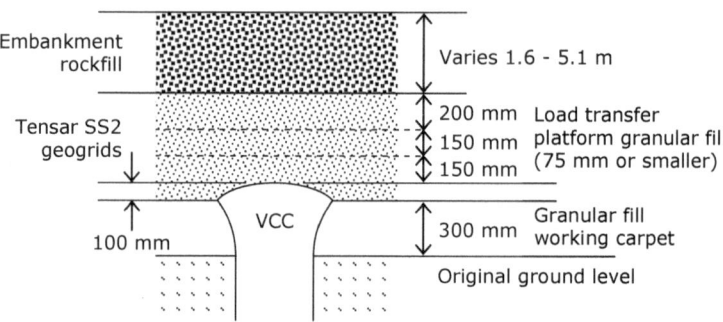

Fig. 6.2 Embankment design for the second Severn crossing

6.6.2 Construction of Apartments on a Site Bordering River Erne, Northern Ireland

A development of 2 and 3 storey town houses and 4 storey apartment blocks were constructed on a site bordering the River Erne in Enniskillen, Northern Ireland during 1999 and 2000. The details of this project were reported in seminar by Milligan (2006).

Ground investigations showed that the underlying subsoil consisted of made ground over substantial depths of peat and soft alluvial clay of thickness of up to 10 m, and underlying glacial till. A (simplified) schematic of the site is shown in Fig. 6.3. A summary of the subsoil properties is shown in Table 6.3. Note that compressibility of the subsoil is very high, for instance compared to the Second Severn Crossing case study.

The site was low-lying and susceptible to flooding so the ground level for the development had to be raised by up to about 3.0 m. Due to the poor ground conditions, a load transfer platform was constructed over the whole area of the site supported by piles into the underlying glacial till (Figs. 6.3 and 6.4). The load transfer platform was used to provide the foundation for the buildings, of conventional construction with shallow strip footings, as well as for all the remainder of the site including gardens, roads and parking areas. It should be noted that there was no direct link between the building footings and piles beneath.

The piles were installed in a triangular arrangement at 2.75 m spacing with a pile cap size of 0.75 m. Beneath the pile caps was 0.5 m thickness of working platform

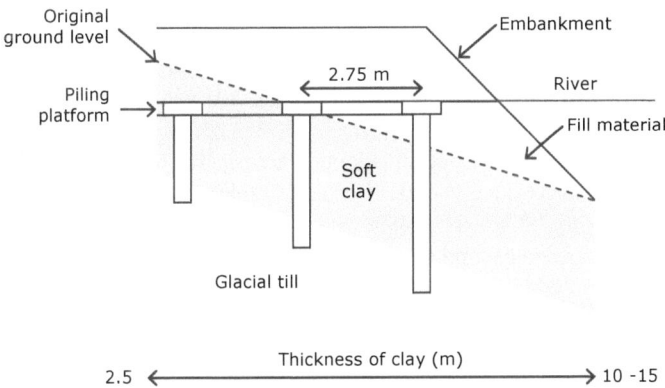

Fig. 6.3 Cross section for the project in Ireland

Table 6.3 Summary of subsoil properties for the project in Ireland		Thickness, t (m)	Stiffness, $E_0(kN/m^2)$
	Alluvial clay	2.5–10	500
	Peat	1–3	200

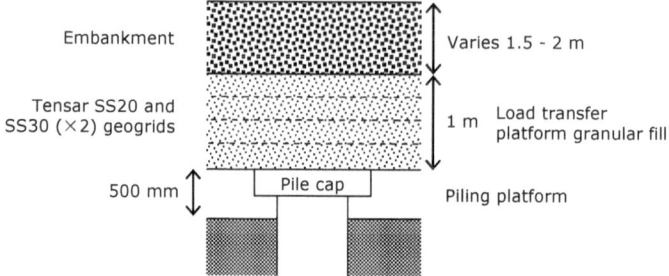

Embankment   Varies 1.5 - 2 m

Tensar SS20 and
SS30 (×2) geogrids   1 m   Load transfer
                                                                          platform granular fill

500 mm   Pile cap   Piling platform

Fig. 6.4 Embankment design for the project in Ireland

Table 6.4 Summary of
geogrid properties for the
project in Ireland

Property	SS20	SS30
Short-term stiffness (kN/m)	280	420

(unsupported fill). Three layers of Tensar geogrid were used; SS20 (× 1) and SS30 (× 2) as shown in Fig. 6.4. The short-term properties of the geogrid are shown in Table 6.4, and again long-term values would be lower.

6.6.3   A650 Bingley Relief Road

The A650, Bingley Relief Road, in West Yorkshire, UK, officially opened to traffic in January 2004. Part of the route involved crossing Bingley North Bog, with soft compressible peat varying in depth up to 11 m (and underlain by glacial sands and gravels), adjacent to a sensitive railway. This projected was reported by Gwede and Horgan (2008).

The piled embankment solution adopted across the North Bog involved the construction of approximately 440 m of low height (2.0 m) piled embankment (varying from 1.8 to 2.2 m). The piles were installed on a square 2.5 m grid, with 900 mm wide square precast pile caps bedded onto the piles. A temporary working platform was installed but subsequently removed once the piles were constructed thus minimising the change in the long-term stress acting on the peat.

No specific information is provided for the compressibility of the subsoil. However, it is indicated that it is very soft and hence it has been assumed to have a one-dimensional modulus of 200 kN/m^2.

The design for the geosynthetic was provided by two orthogonal (longitudinal and transverse) layers of Stabilenka 600/50. Based on information in Gwede and Horgan (2008), the long-term stiffness of the geosynthetic including creep was deduced from a load/strain plot as 4800 kN/m.

6.6.4   A1/N1 Flurry Bog

A section of the A1/N1 dual carriageway between Dundalk and Newry forming the cross border link between the Republic of Ireland and Northern Ireland has recently been completed (August 2007). This project was reported by Orsmond (2008).

The Flurry Bog is a cutaway bog where the levels had been reduced to groundwater level with poor drainage exasperated by the adjacent Salmonoid River that is prone to flooding. Ground conditions were typically very weak peat 4–6 m deep overlying soft silt 3–4 m, then a thin layer of gravel over bedrock.

The embankment height was about 3.0 m over the piles. The pile cap size was 0.8 m and pile spacing was 2.5 m on a square grid. A working platform was constructed, comprising of two layers of geogrid spaced within 600 mm rockfill, which proved sufficient for the intended loads but weak enough for the piles to be driven through it. Again, no specific information is provided for the compressibility of the subsoil, but again it appears to be very soft, and a one-dimensional modulus of 200 kN/m^2 has been assumed.

The final design incorporated Ployfelt PET woven polyester geosynthetic laid in longitudinal and transverse directions, with strength varying between 540 and 780 kN/m. Based on manufacturers literature for these products a typical long-term stiffness was taken as 5000 kN/m.

6.6.5   Case Study Comparison

The project at the Second Severn Crossing has been considered as a successful case, in which the settlement (both absolute and differential) of the embankment have remained within acceptable limits. However, the project in Northern Ireland was not. Within two years of completion, the ground deformations around the buildings constructed on the LTP were becoming noticeable. Some time later, the pile caps 'punched' into the material above, causing significant deformation. According to detailed investigation and assessment of the cause of failure (Milligan, 2006), the problems were caused by excessive and continuing deformation of the load transfer platform. Equation (6.4) will be used to consider these four cases, as summarised in Table 6.5.

The pile cap spacing (s) shows relatively little variation between the cases. However, the pile cap size (a) is slightly smaller at the Second Severn Crossing, where enlarged heads on the VCC were used rather than actual pile caps. The first two case studies consider a triangular grid of piles, but they will be treated as if the grid was square using Eq. (6.4).

The height of the embankments (h) are not that significant, generally corresponding to about 1.5 times the clear spacing ($s - a$). On the A650 h_w is zero since the precaution was taken of removing the working platform.

Table 6.5 Summary of input parameters

	Second Severn crossing			Construction of apartments, Ireland		A650	A1/N1
Piled embankment geometry							
s(m)	2.7			2.75		2.5	2.5
a(m)	0.5			0.75		0.9	0.8
h(m)	3.5			3.0		2.0	3.0
h_w(m)	0.3			0.5		0.0	0.6
Subsoil properties							
E_0(kN/m^2)	5000	1800	500	500	200	200	200
h_s	1.5	2.5	4	7	3	10	10
J (kN/m)	150	150	140	210	210	4800	5000
Derived parameters							
Clear spacing: $(s - a)$	2.2			2.0		1.6	1.7
Dimensionless working platform thickness: $h_w/(s - a)$	0.14			0.25		0	0.35
Dimensionless subsoil factor: $E_0/\gamma h_s$	6.07			2.03		1.18	1.18
Dimensionless subsoil factor: $5J/\gamma(s - a)^2$	18.23			41.18		551.47	508.85

For various layers of soft subsoil, the total settlement for a given stress can be calculated by:

$$\frac{\delta_s}{\sigma_s} = \frac{h_1}{E_{01}} + \frac{h_2}{E_{02}} + \dots \frac{h_n}{E_{0n}}$$ (6.5)

Hence for the purposes of defining the variables in Eq. (6.4) the subsoil thickness is taken as the sum of thicknesses for all soft layers and then a representative stiffness can be derived as follows:

$$\frac{E_0}{h_s} = \frac{1}{h_1/E_{01} + h_2/E_{02} + \dots h_n/E_{0n}}$$ (6.6)

The subsoil thickness for the Second Severn Crossing and apartments in Ireland are based on typical values for the least soft layers, but maximum values for the softest layer, giving a 'worst case scenario'. For the other cases a nominally very low stiffness was used. Notably the derived 'dimensionless subsoil factor' is highest by a significant margin for the Second Severn Crossing.

The equation only considers the reinforcement contribution due to membrane tension, and not any other interaction with the soil. Thus the effect of multiple layers of biaxial geogrids is incorporated simply by assuming all grids to deform by the same amount and summing the stiffness (J) of the various grids. Where geosynthetics have been laid in orthogonal directions they have effectively been treated as a single biaxial grid with the maximum stiffness in each direction. An approximate long-term stiffness has been assumed throughout.

For the Second Severn Crossing a total biaxial stiffness of 300 kN/m has been assumed, based on two layers of geogrid with short-term stiffness of 150 and 300 kN/m in orthogonal directions. Thus the long-term value is assumed as two-thirds of the average stiffness (kN/m):

$$300 = 2 \times 0.67 \times (300 + 150)/2.$$

For the apartments in Ireland where large deformations were known to have occurred the long-term stiffness was taken as half the nominal short-term stiffness at normal working strain, giving a total stiffness of 560 kN/m.

The remaining case studies used two orthogonal layers of geosynthetic, so the biaxial stiffness has been taken as the long-term stiffness of one layer. It is evident that the reinforcement stiffness is much higher in these cases.

The remaining terms in Eq. (6.4) are the normalised working platform thickness (as summarised in Table 6.5), and A. Since the embankments are not that high (and punching of the pile caps into the base of the embankment is unlikely to be an issue) A has been taken as 0.5 (Sect. 6.4).

Table 6.6 shows the numerical solution of Eq. (6.4), with results for the compatible displacement at equilibrium and corresponding reinforcement strain, and the distribution of load between the geogrid and subsoil. Figure 6.5 shows the corresponding interaction diagrams. Note that the point of maximum arching for the Ground Reaction Curve is assumed to extend from 2% normalised displacement at a constant value. In reality some form of brittle response would be observed (Sect. 2.3, Fig. 2.15), but this is not considered in Eq. (6.4). Equilibrium is satisfied when the sum of the subsoil and geogrid response meets the point of maximum arching, $[A + h_w/(s - a)]$.

Table 6.6 Summary of results

	Second Severn crossing	Construction of apartments, Ireland	A650	A1/N1
Deformation				
$\delta/(s - a)$ (%)	10.2	20.2	8.9	11.2
δ (mm)	224.4	404.0	142.4	190.4
Geogrid strain ε (%)	2.77	10.88	2.11	3.35
Load distribution				
Load distribution	23.8	25.5	13.6	24.6
Total stress (kN/m^2)	23	14	2.7	3.8
Subsoil stress (kN/m^2)	0.8	11.5	10.9	20.8

Table 6.6 shows that the deformation is about twice as large for the Irish apartments as any of the other cases, and the geogrid strain is over 10%, compared with less than 4% in the other cases. This is consistent with the observation that failure of the LTP was observed for the Irish apartments. At the Second Severn Crossing, where

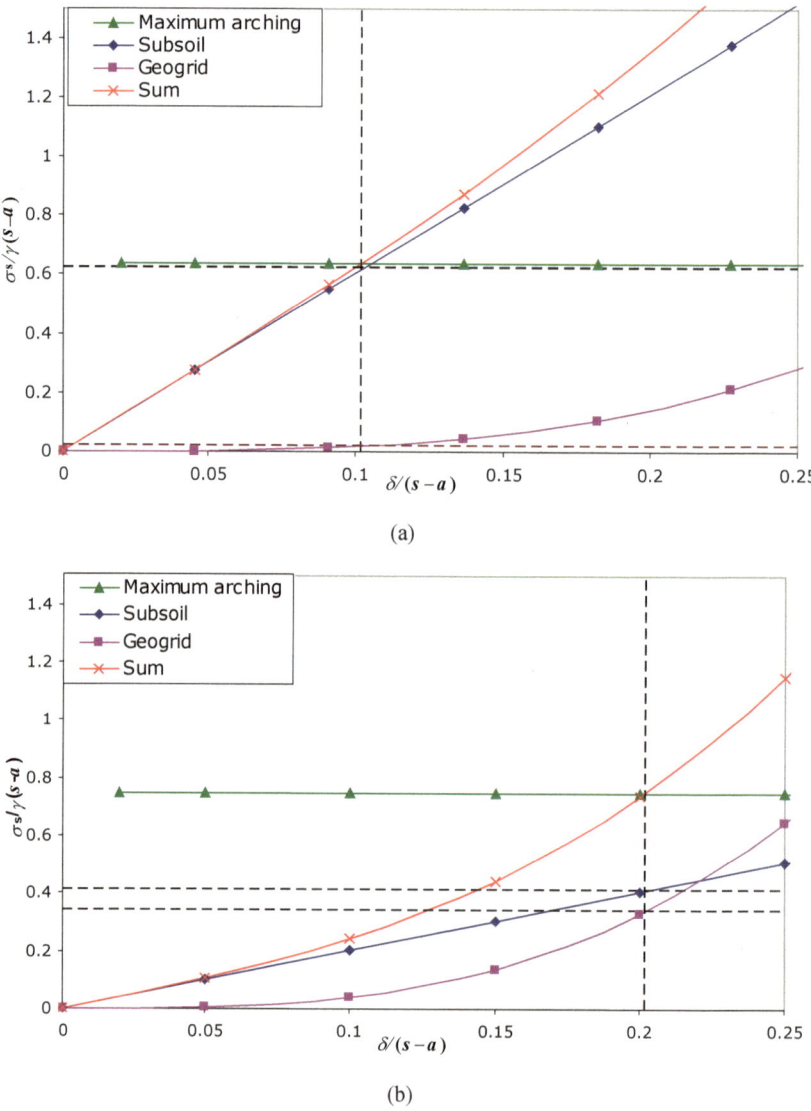

(a)

(b)

Fig. 6.5 Interaction diagrams; **a** second Severn crossing, **b** construction of apartments, Ireland, **c** A650, **d** A1/N1

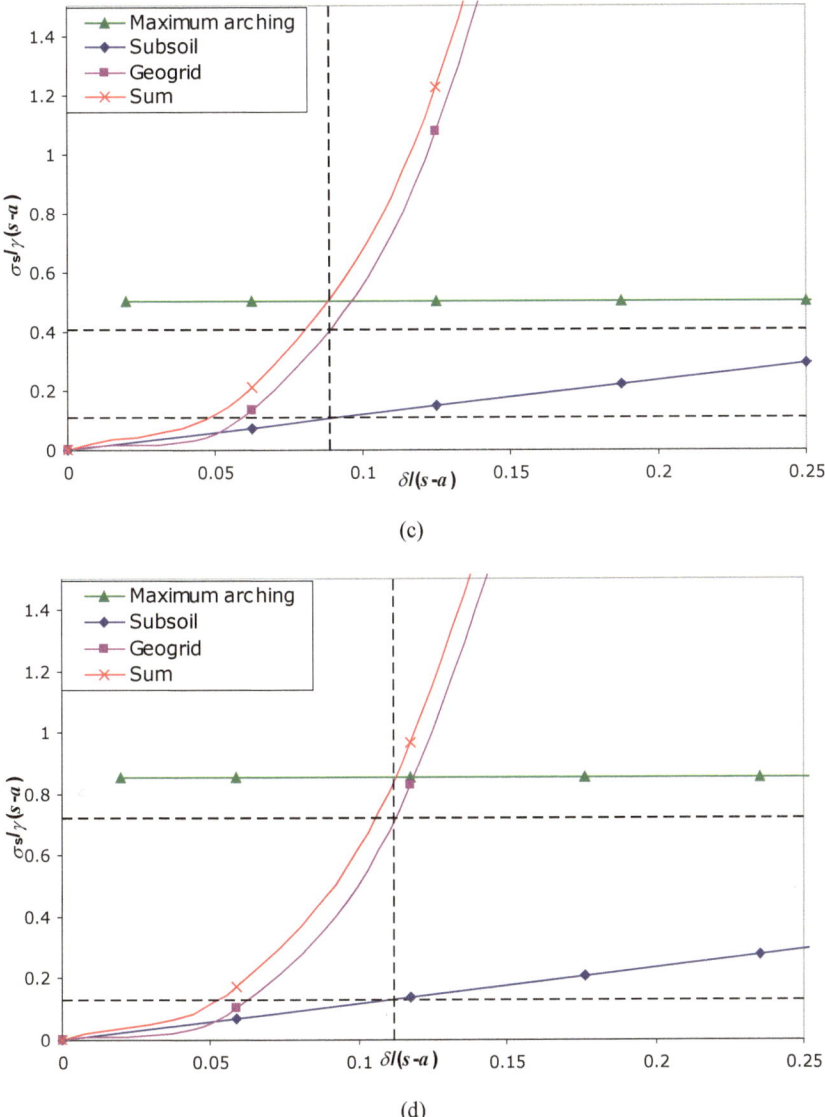

Fig. 6.5 (continued)

reinforcement stiffness is low, the subsoil (which is relatively competent) carries most of the load, whereas for the A650 and A1/N1 this situation is reversed.

Solution of the equation does not consider the condition that the geogrid cannot carry a normalised load exceeding *A* (from the embankment), and likewise that the subsoil must carry the load from the working platform as a minimum. These conditions are most likely to be encountered when the geogrid is stiff and the subsoil is very compressible. The situation does not arise for the first two case studies (where the geogrid is not that stiff), or the third (where there is no working platform). However, it does occur for the A1/N1. Here the geogrid would carry only the embankment arching load:

$$5\frac{J}{\gamma(s-a)^2}\left(\frac{\delta}{s-a}\right)^3 = A \tag{6.7}$$

and hence $\delta/(s-a) = (0.5/509)^{1/3} = 9.9\%$, with corresponding reinforcement strain 2.6%—somewhat less than in Table 6.6.

Likewise for the subsoil:

$$\frac{E_s}{\gamma h_s}\frac{\delta}{s-a} = \frac{h_w}{s-a} \tag{6.8}$$

and hence $\delta/(s-a) = (0.35/1.18) = 30\%$.

As anticipated this indicates that the subsoil settles more than the sag in the reinforcement, and hence a gap is formed below the reinforcement. However, it is worth reflecting that the subsoil properties assumed for this case are probably conservative and thus this may not actually happen.

6.7 Summary

The ratio $h/(s - a)$ has been considered as an important parameter for aching behaviour in design. It has been demonstrated that the critical value of $h/s = 1.5$ for full arching reported in this work corresponds to $h/(s - a) \approx 0.2 \sim 0.3$ which was consistent with the values reported by the Aslam (2008), Ellis and Aslam (2009a, 2009b), and Potts and Zdravkovic (2008).

The value of $\sigma_s/\gamma(s - a)$ at the point of maximum arching for medium height embankments was between 0.4 and 0.8 in this study, which was again consistent with the values reported by the other authors for physical and numeric modelling.

An interaction diagram has been described (following the concept proposed by Ellis & Aslam, 2009b), where the combined action of arching in the embankment, reinforcement membrane action and the subsoil give equilibrium at a compatible settlement. Equation (6.4) is based on the interaction diagram, using terms for 3D behaviour derived in Chaps. 4 and 5.

The interaction diagram and accompanying equation were applied to a number of case studies: two using a 'Load Transfer Platform' (LTP) with multiple layers of low stiffness/strength geogrid, and two with geotextile reinforcement which was approximately 10 times stiffer. Case studies often contain relatively limited information regarding the subsoil, which is very soft, and has not been extensively considered in the design. In the case studies with relatively stiff reinforcement it was (perhaps conservatively) assumed that the subsoil was very soft. However, the geotextile was able to carry the arching embankment load at a tolerable strain.

In the case studies for the LTPs there was information regarding the subsoil compressibility. In one case the subsoil was soft, but not extremely soft, and here it was found that the subsoil actually carried a significant portion of the arching embankment and working platform load, significantly reducing the load on the geogrid. However, in the other case a combination of very soft subsoil, and geogrid with relatively low stiffness implied intolerable geogrid strain, correctly reflecting an actual failure in the field.

Thus the question regarding design of LTPs with low strength geogrids according to the 'Guido' method still remains. This design method is acknowledged (e.g. Jenner et al., 1998) to fundamentally differ from an approach based on catenary action of the reinforcement. Equation (6.4) is based on catenary action, and has been verified by the FE analyses in this monograph without indication that it is inappropriate.

References

Aslam, R., Ellis, E. A. (2008). Centrifuge modelling of piled embankments. In *Proceedings of ISSMGE 1st international conference on transportation geotechnics*, Nottingham, UK. Taylor and Francis Group, London, ISBN 978-0-415-47590-7, pp. 363–368.

Abusharar, S. W., Zheng, J. J., Chen, B. G., & Yin, J. H. (2009). A simplified method for analysis of a piled embankment reinforced with geosynthetics. *Geotextiles and Geomembranes, 27*, 39–52.

Ellis, E. A., & Aslam, R. (2009a). Arching in piled embankments: Comparison of centrifuge tests and predictive methods—Part 1 of 2. In *Ground engineering* (pp. 34–38), June 2009.

Ellis, E. A., & Aslam, R. (2009b). Arching in piled embankments: Comparison of centrifuge tests and predictive methods—Part 2 of 2. In *Ground engineering* (pp. 28–31), July 2009.

Ellis, E. A., Zhuang, Y., Aslam, R., & Yu, H. S. (2009). Discussion of 'a simplified method for analysis of a piled embankment reinforced with geosynthetics.' *Geotextiles and Geomembranes, 27*, 39–52.

Gwede, D., & Horgan, G. J. (2008). Design, construction and in-service performance of a low height geosynthetic reinforced piled embankment: A650 Bingley Relief Road. In *Proceedings of the 4th European geosynthetics conference*, Edinburgh, UK, September 2008, Paper number 256.

Jenner, C. G., Austin, R. A., & Buckland, D. (1998). Embankment support over piles using geogrids. In *Proceedings 6th international conference on geosynthetics*, Atlanta, USA, pp. 763–766.

Maddison, J. D., Jones, D. B., Bell, A. L., & Jenner, C. G. (1996). Design and performance of an embankment supported using low strength geogrids and vibro concrete columns. In *Geosynthetics—Applications, design and construction*, De Groot, Den Hoedt and Termaat, pp. 325–332.

Milligan, G. (2006). Seminar given at the ICE setting out the findings and conclusions from a detailed investigation and assessment of the causes of failure of apartment blocks constructed on a site bordering the River Erne in Enniskillen, Northern Ireland.

Orsmond, W. (2008). A1N1 flurry bog piled embankment design, construction and monitoring. In *Proceedings of the 4th European geosynthetics conference*, Edinburgh, UK, September 2008, Paper number 290.

Potts, V. J., & Zdravkovic, L. (2008). Finite element analysis of arching behaviour in soils. In *The 12th international conference of international association for computer methods and advances in geomechanics (IACMAG)*, Goa, India, October, 2008, pp. 3642–3649.

Russell, D., & Pierpoint, N. (1997). An assessment of design methods for piled embankments. In *Ground engineering* (pp. 39–44), November 1997.

Chapter 7
Conclusions

Numerical modelling of arching in piled embankments has been undertaken in this monograph. The research has improved generic understanding of arching behaviour, and interaction of the embankment with the subsoil and any layers of geogrid or geosynthetic reinforcement placed at the base of the embankment. The improved understanding of behaviour highlights the inadequacies in some existing design approaches, and has been used to develop a simple equation for use in design.

The numerical analyses focus on the unreinforced piled embankments without subsoil in plane strain and 3D situations respectively. In both cases at $(h/s) \approx 0.5$, there is no evidence of arching based on the stress acting on the subsoil, and very large differential settlement at the embankment surface occurs. As (h/s) increases to 1.5 the stress on the subsoil does not increase significantly (and thus there is significant evidence of arching), and differential settlement at the embankment surface tends to zero. The value of $\sigma_s/\gamma s$ for maximum arching is approximately 0.25–0.40 except for high s/a in the 3D situation. At higher values of (h/s) conditions at the pile cap are critical, and Eq. (3.2) can be used to conservatively estimate the stress on the subsoil. The vertical extent of arching in the embankment was found to be 1.25–1.50 times the centre-to-centre pile spacing. This facilitates the determination of the minimum height of a piled embankment for effective arching.

As for the geogrid reinforced piled embankment, the results showed that with the effect of geogrid, the ultimate stress on the subsoil can be reduced to zero. However, this required significant sag of the geogrid reinforcement. The Eq. (2.30) (based on a parabola) can be used to predict the geogrid action for both plane strain and 3D situations (using an appropriate span). The plane strain and 3D analyses also found that the sag of reinforcement was very sensitive to the span of the reinforcement between piles, but relatively insensitive to its stiffness. For the embankment with three layers of geogrid, the upper two layers of grids showed less settlement compared to bottom layer, and thus less tension. It was also found that the tension in the reinforcement was approximately constant across the span in plane strain analyses.

© The Author(s) 2025 131
Y. Zhuang, *Arching in Piled Embankments Including the Effects of Reinforcement and Subsoil*, https://doi.org/10.1007/978-981-96-1198-0_7

However, in the 3D situation, the maximum tension in the geogrid occurred at the corner of the pile cap.

This monograph also presented predictions by BS 8006, the 'Marston' and Hewlett and Randolph (1988) approaches for a wide range of piled embankment geometries, and comparison with results from three-dimensional Finite Element (FE) analyses. A suggested modification to the BS 8006 formula for prediction of reinforcement sag gives good agreement with the results obtained from FE analyses. There are arguments that the component Equations under- and over-predict various aspects of the behaviour. Nevertheless, the ultimate prediction of T_{rp} (including modification for subsoil support) is quite good compared to the FE data. The BS 8006 (2012) modified prediction is 'best' (but sometimes slightly unconservative), whilst the BS 8006 (2010) modified prediction is conservative in all cases considered.

The reinforced piled embankment with subsoil in plane strain and 3D situations were presented in this monograph. The analyses showed that the subsoil could give a major contribution to overall vertical equilibrium. In fact, the contribution from the subsoil exceeded a prediction based on simple 1D settlement, due to the effect of principle stress rotation (effect of bearing) near the top of the subsoil. A simple equation (or interaction diagram) based on the results has been proposed to enable assessment of the relative contribution of the reinforcement and subsoil to the equilibrium, and hence to predict the load and strain in the reinforcement.

References

BS 8006-1. (2010). Code of practice for strengthened/reinforced soils and other fills. British Standards Institution.

BS 8006-1. (2012). Code of practice for strengthened/reinforced soils and other fills, incorporating Corrigendum.

Hewlett, W. J., & Randolph, M. F. (1988). Analysis of piled embankments. In Ground engineering (12–18), April 1988.